Using SAS for Data Management, Statistical Analysis, and Graphics

Using SAS for Data Management, Statistical Analysis, and Graphics

Ken Kleinman

Department of Population Medicine

Harvard Medical School

Boston, Massachusetts

Nicholas J. Horton

Department of Mathematics and Statistics

Smith College

Northampton, Massachusetts

CRC Press is an imprint of the
Taylor & Francis Group an **informa** business
A CHAPMAN & HALL BOOK

CRC Press
Taylor & Francis Group
6000 Broken Sound Parkway NW, Suite 300
Boca Raton, FL 33487-2742

Printed in the United States of America on acid-free paper
10 9 8 7 6 5 4 3 2

International Standard Book Number: 978-1-4398-2757-4 (Paperback)

Library of Congress Cataloging-in-Publication Data

Kleinman, Ken.
Using SAS for data management, statistical analysis, and graphics / Ken Kleinman, Nick Horton.
p. cm.
Includes bibliographical references and index.
ISBN 978-1-4398-2757-4 (pbk. : alk. paper)
1. SAS (Computer program language) 2. Database management. 3. Mathematical statistics--Data processing. I. Horton, Nicholas J. II. Title.

QA76.73.S27K54 2011
005.3--dc22 2010021410

Visit the Taylor & Francis Web site at
http://www.taylorandfrancis.com

and the CRC Press Web site at
http://www.crcpress.com

Contents

List of Tables

List of Figures

Preface

SAS™ (SAS Institute, 2009) is a statistical software package used in many fields of research. We have written this book as a reference text for users of SAS. Our primary goal is to provide users with an easy way to learn how to perform an analytic task, without having to navigate through the extensive, idiosyncratic, and sometimes unwieldy documentation provided. We include many common tasks, including data management, descriptive summaries, inferential procedures, regression analysis, multivariate methods, and the creation of graphics. We also show a small sample of the many more complex applications available. In toto, we hope that the text will serve as a brief summary of the features of SAS most often used by statistical analysts.

We do not attempt to exhaustively detail all possible ways available to accomplish a given task. Neither do we claim to provide the most elegant solution. We have tried to provide a simple approach that is easy for a new user to understand, and have supplied several solutions when this seems likely to be helpful.

Who should use this book?

Those with an understanding of statistics at the level of multiple-regression analysis will find this book helpful. This group includes professional analysts who use statistical packages almost every day as well as statisticians, epidemiologists, economists, engineers, physicians, sociologists, and others engaged in research or data analysis. We anticipate that the book will be particularly useful for relatively new users, those with some experience using SAS but who often find themselves frustrated by the documentation provided. However, even expert SAS programmers may find it valuable as a source of task-oriented, as opposed to procedure-oriented, information. In addition, the book will bolster the analytic abilities of a new user of SAS, by providing a concise reference manual and annotated examples.

Using the book

The book has two indices, in addition to the comprehensive "Table of Contents." The first index is organized by topic (subject), in English; the other is a SAS index, organized by SAS syntax. You can use the "SAS Index" to look up a task for which you know the approximate SAS keywords but need to find the exact syntax or an example. You can use the "Subject Index" to find syntax or applications when the syntax is unknown.

Extensive example analyses are presented. Table A.1 is a comprehensive list of example applications employing data from the HELP study, which is described in the Appendix. Additional case studies, usually with more complex coding, are shown in Chapter 7. Readers are encouraged to download the dataset and code from the book Web site. The examples demonstrate the code in action and facilitate exploration by the reader. In the indices, example applications are listed with *italicized* page numbers.

Where to begin

We do not anticipate that the book will be read cover to cover. Instead, we hope that the extensive indexing, cross-referencing, and worked examples will make it possible for readers to directly find and then implement what they need. A user new to SAS should begin by reading Chapter 1, which includes a sample session and overview. Other users may want to skip to the indices or table of contents.

On the Web

The book Web site at `http://www.math.smith.edu/sas` includes the "Table of Contents," the indices, the HELP dataset, example code in SAS, and a list of erratum.

Acknowledgments

We would like to thank Rob Calver, Kari Budyk, Shashi Kumar, Sarah Morris, and Linda Leggio for their support and guidance at Informa CRC/Chapman & Hall. We also thank Allyson Abrams, Russell Lenth, Brian McArdle, Richard Platt, and David Schoenfeld for contributions to SAS or LaTeX programming efforts, comments, guidance, support, and/or helpful suggestions on drafts of the manuscript.

Above all we greatly appreciate Sara and Julia as well as Abby, Alana, Kinari, and Sam, for their patience and support.

Amherst, Massachusetts and Northampton, Massachusetts

Chapter 1

Introduction to SAS

The SAS™ system is a programming and data analysis package developed and sold by SAS Institute, Cary, North Carolina. SAS markets many products; when installed together they result in an integrated environment. In this book we address software available in the Base SAS, SAS/STAT, SAS/GRAPH, SAS/ETS, and SAS/IML products. Base SAS provides a wide range of data management and analysis tools, while SAS/STAT and SAS/GRAPH provide support for more sophisticated statistical methods and graphics, respectively. We touch briefly on the IML (interactive matrix language) module, which provides extensive matrix functions and manipulation, and the ETS (Econometrics and Time Series) module, which supports time series tools and other specialized procedures.

SAS Institute also markets some products at reduced prices for individuals as well as for educational users. The "Learning Edition" lists at $199 as of March 2009, but limits use to only 1,500 observations (rows in a dataset). More information can be found at `http://support.sas.com/learn/le/order.html`. Another option is SAS "OnDemand for Academics" (`http://www.sas.com/govedu/edu/programs/oda_account.html`) currently free for faculty and $60 for students. This option uses servers at SAS to run code and has a slightly more complex interface than the local installation discussed in this book.

1.1 Installation

SAS products are available for a yearly license fee. Once licensed, instructions for download are provided; this includes detailed installation instructions tailored to the operating system for which the license was obtained. Also necessary is a special "setinit" file which functions as a password allowing installation of licensed products. An updated setinit file is provided upon purchase of a license renewal.

1.2 Running SAS and a sample session

Once installed, a recommended step for a new user is to start SAS and run a sample session. Starting SAS in a GUI environment opens a SAS window as displayed in Figure 1.1.

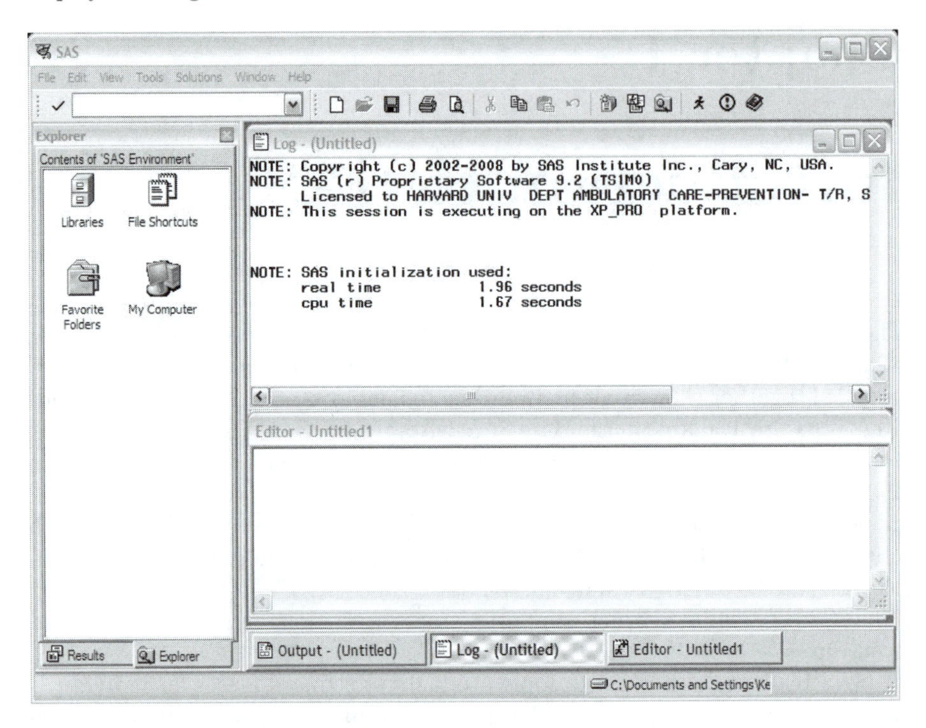

Figure 1.1: The SAS Windows interface.

The window is divided into two panes. On the left is a navigation pane with Results and Explorer tabs, while the right is an interactive windowing environment with Editor, Log, and Output Windows. Effectively, the right-hand pane is like a miniature graphical user interface (GUI) in itself. There are multiple windows, any one of which may be maximized, minimized, or closed. In Figure 1.1 the right-hand pane displays the Log and Editor windows. The window contents can also be saved to the operating system or printed. Depending on the code submitted, additional windows may open in this area. To open a window, click on its name at the bottom of the right-hand pane; to maximize or minimize within the SAS GUI, click on the standard icons your operating system uses for these actions.

On starting SAS, the cursor will appear in the Editor window. Commands such as those in the sample session which follows are typed there. They can also be read into the window from previously saved text files using *File; Open*

Program from the menu bar. Typing the code does not do anything, even if there are carriage returns in it. To run code, it must be *submitted* to SAS; this is done by clicking the submit button in the GUI as in Figure 1.2 or using keyboard shortcuts. After code is submitted SAS processes the code. Results are not displayed in the Editor window, but in the Output window, and comments from SAS on the commands which were run are displayed in the Log window. If output lines (typically analytic results) are generated, the Output window will jump to the front.

In the left-hand pane, the Explorer tab can be used to display datasets created within the current SAS session or found in the operating system. The datasets are displayed in a spreadsheet-like format. Navigation within the Explorer pane uses idioms familiar to users of GUI-based operating systems. The Results tab allows users to navigate among the output generated during the current SAS session. The Explorer and Results panes can each be helpful in reviewing data and results, respectively.

As a sample session, consider the following SAS code, which generates 100 normal variates (see Section 2.10.5) and 100 uniform variates (see Section 2.10.3), displays the first five of each (see Section 2.2.4), and calculates series of summary statistics (see Section 3.1.1). These commands would be typed directly into the Editor window:

```
data test;
  do i = 1 to 100;
    x1 = normal(0);
    x2 = uniform(0);
    output;
  end;
run;

proc print data=test (obs=5);
run;

ods select moments;
proc univariate data=test;
  var x1 x2;
run;
```

A user can run a section of code by selecting it using the mouse and clicking the "running figure" (submit) icon near the right end of the toolbar as shown in Figure 1.2. Clicking the submit button when no text is selected will run all of the contents of the window. This code is available for download from the book Web site: `http://www.math.smith.edu/sas/examples/sampsess.sas`.

We discuss each block of code in the example to highlight what is happening.

Figure 1.2: Running a program.

```
data test;
  do i = 1 to 100;
    x1 = normal(0);
    x2 = uniform(0);
    output;
  end;
run;
```

After selecting and submitting the above code the Output window will be empty, since no output was requested, but the log window will contain some new information:

```
1    data test;
2      do i = 1 to 100;
3        x1 = normal(0);
4        x2 = uniform(0);
5        output;
6      end;
7    run;

NOTE: The dataset WORK.TEST has 100 observations and 3 variables.
NOTE: DATA statement used (Total process time):
      real time               0.01 seconds
      cpu time                0.01 seconds
```

This indicates that the commands ran without incident, creating a dataset called `WORK.TEST` with 100 rows and three columns (one for `i`, one for `x1`, and one for `x2`). The line numbers can be used in debugging code.

Next consider the `proc print` code.

```
proc print data=test (obs=5);
run;
```

When these commands are submitted, SAS will generate the following in the Output window. Note that only 5 observations are shown because `obs=5` was specified (1.6.1). Omitting it will cause all 100 lines of data to be printed.

```
Obs    i          x1           x2

  1    1    -1.50129      0.80471
  2    2    -0.79413      0.96266
  3    3     0.16263      0.71409
  4    4     0.01924      0.90480
  5    5     0.11481      0.31984
```

Data are summarized by submitting the lines specifying the `univariate` procedure.

```
ods select moments;
proc univariate data=test;
  var x1 x2;
run;
ods select all;
```

```
The UNIVARIATE Procedure
Variable:  x1

N                          100    Sum Weights               100
Mean                -0.0536071    Sum Observations   -5.3607128
Std Deviation       1.08284966    Variance           1.17256338
Skewness            0.34607815    Kurtosis           0.81876938
Uncorrected SS      116.371147    Corrected SS       116.083775
Coeff Variation     -2019.9733    Std Error Mean     0.10828497
```

```
Variable:  x2

N                          100    Sum Weights               100
Mean                0.46907241    Sum Observations   46.9072407
Std Deviation       0.28927737    Variance            0.0836814
Skewness            0.21045221    Kurtosis           -1.0485607
Uncorrected SS      30.2873506    Corrected SS       8.28445825
Coeff Variation     61.6700887    Std Error Mean     0.02892774
```

Similar to the `obs=5` specified in the `proc print` statement above, the `ods select moments` statement causes only a subset of the output to be printed. By default, SAS often generates voluminous output that can be hard for new users to digest and would take up many pages of a book. We use the `ODS` system (1.7) to select pieces of the output throughout the book.

For each of these submissions, additional information is presented in the Log window. While some users may ignore the Log window unless the code did not work as desired, it is always a good practice to examine the log carefully, as it contains warnings about unexpected behavior as well as descriptions of errors which cause the code to execute incorrectly or not at all.

Note that the contents of the Editor, Log, and Output windows can be saved in typical GUI fashion by bringing the window to the front and using *File; Save* through the menus.

Figure 1.3 shows the appearance of the SAS window after running the sample program. The Output window can be scrolled through to find results, or the Results tab shown in the left-hand pane can be used to find particular pieces of output more quickly. Figure 1.4 shows the view of the dataset found through the Explorer window by clicking through *Libraries; Work; Test.* Datasets not assigned to permanent storage in the operating system (see writing native files, 2.2.1) are kept in a temporary library called the "Work" library.

1.3 Learning SAS and getting help

There are numerous tools available for learning SAS, of which at least two are built into the program. Under the Help menu in the Menu bar are "Getting Started with SAS Software" and "Learning SAS Programming." In the online

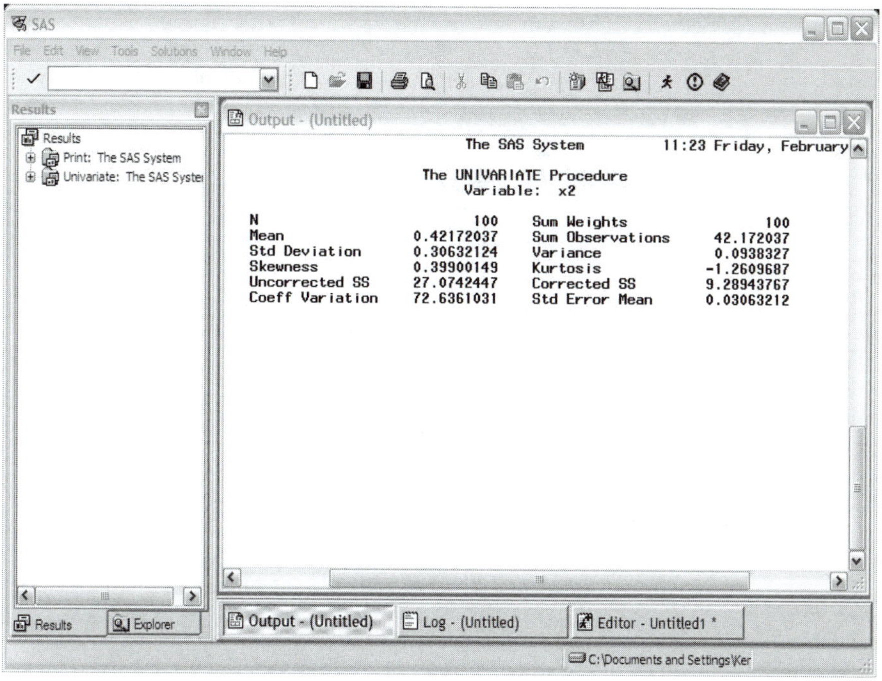

Figure 1.3: The SAS window after running the sample session code.

help, under the `Contents` tab is "Learning to Use SAS" with many entries included. For those interested in learning about SAS but without access to a working version, some Internet options include the excellent University of California–Los Angeles (UCLA) statistics Web site, which includes the "SAS Starter Kit" (`http://www.ats.ucla.edu/stat/sas/sk/default.htm`). While dated, the slide show available from the Oregon State University Statistics Department could be useful (see `http://oregonstate.edu/dept/statistics/software/sas/2002seminar`). The SAS Institute offers several ways to get help. The central place to start is its Web site where the front page for support is `http://support.sas.com/techsup`, which has links to discussion forums, support documents, and instructions for submitting an e-mail or phone request for technical support.

Complete documentation is included with SAS installation by default. Clicking the icon of the book with a question mark in the GUI (Figure 1.5) will open a new window with a tool for viewing the documentation (Figure 1.6). While there are `Contents`, `Index`, `Search`, and `Favorites` tabs in the help tool, we generally use the `Contents` tab as a starting point. Expanding the `SAS Products` folder here will open a list of SAS packages (Base SAS, SAS/STAT, etc.). Detailed documentation for the desired procedure can be found under

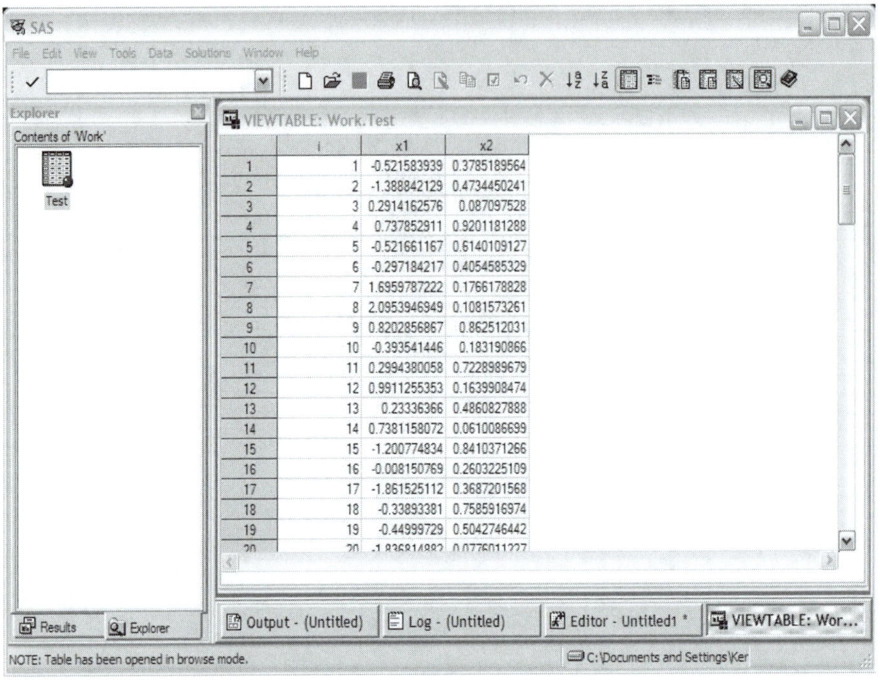

Figure 1.4: The Explorer window.

the package which provides access to that `proc` or, as of SAS 9.2, in the alphabetical list of procedures found in: Contents; SAS Products; SAS Procedures. In the text, we provide occasional pointers to the online help, using the folder structure of the help tool to provide directions to these documents. Our pointers use the SAS 9.2 structure; earlier versions have a similar structure except that procedures must be located through their module. For example, to find the `proc mixed` documentation in SAS 9.2, you can use: Contents; SAS Products; SAS Procedures; MIXED, while before version 9.2, you would navigate to: Contents; SAS Products; SAS/STAT; SAS/STAT User's Guide; The MIXED Procedure.

1.4 Fundamental structures: Data step, procedures, and global statements

Use of SAS can be broken into three main parts: the data step, procedures, and global statements. The data step is used to manage and manipulate data. Procedures are generally ways to do some kind of analysis and get results. Users of SAS refer to procedures as "procs." Global statements are generally used to

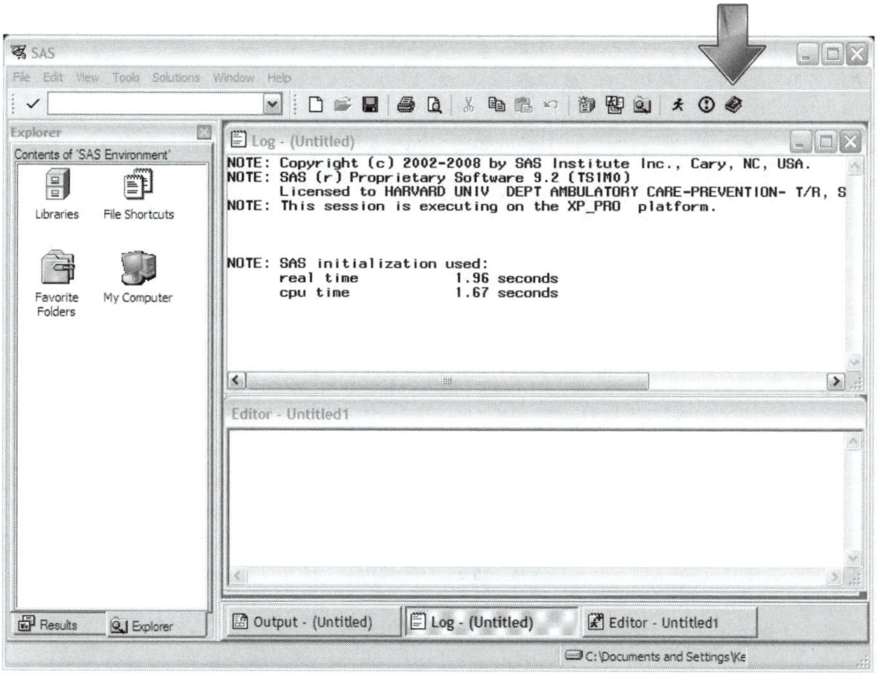

Figure 1.5: Opening the online help.

set parameters and make optional choices that apply to the output of one or more procedures.

A typical data step might read as follows.

```
data newtest;
set test;
  logx = log(x);
run;
```

In this code a new variable named `logx` is created by taking the natural log of the variable `x`. The `data` step works by applying the instructions listed, sequentially, to each line of the dataset named using the `set` statement, then writing that line of data out to the dataset named in the `data` statement. Data steps and procedures are typically multistatement collections. Both are terminated with a `run` statement. As shown above, statements in SAS are separated by semicolons, meaning that carriage returns and line breaks are ignored. When SAS reads the `run` statement in the example (when it reaches the "`;`" after the word `run`), it writes out the processed line of data, then repeats for each line of data until it reaches the end of the `set` dataset. In this example, a line of data is read from the `test` dataset, the `logx` variable is generated,

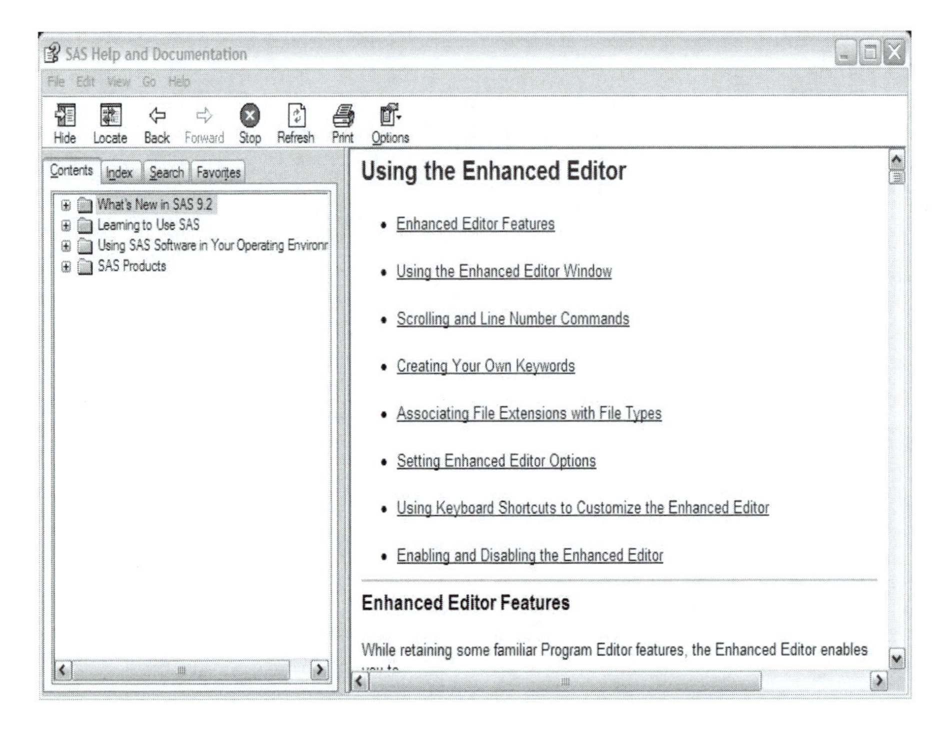

Figure 1.6: The Help and Documentation window.

and the line of data (including `logx`, `x`, and any other data stored in `test`) is written to the new dataset `newtest`.

A typical procedure in SAS might read as follows.

```
proc glm data=newtest;
  model y = logx / solution;
run;
```

Many procedures require multiple statements to function. For example, the `glm` procedure requires both a `proc glm` statement and a `model` statement.

Here, we show the two ways that *options* can be specified in SAS. One way is by simply listing optional syntax after the statement name. In the `proc glm` statement above, we specify, using the `data` option, that the dataset that should be used is the `newtest` dataset. Without this option SAS defaults to using the most recently created dataset. As a matter of good programming practice, we always specify the dataset using the `data` option, which can be used with any and all `procs`. Naming datasets explicitly in each procedure minimizes errors and makes code clearer.

The `model` statement shown demonstrates another way that options are specified, namely after a forward slash. In general, this syntax is used when the main body of the statement may include separate words. For example, the slash in the `model` statement above separates the model specification from the options (here the `solution` option requests the parameter estimates in addition to the default analysis of variance [ANOVA] table). See Section 4.1.1 for more on `proc glm`.

We refer to any SAS code appearing between semicolons generically as "statements." Most statements appear within data steps or procs. Global statements are special statements that need not appear within a data step or a proc. An example would be the following code.

```
options ls=64 ps=60 nocenter;
```

This `options` statement affects the formatting of output pages, limiting the line length to 64 characters per line for 60 lines per page, while removing the default centering.

1.5 Work process: The cognitive style of SAS

A typical SAS work session involves first writing a `data` step or loading a saved command file (conventionally saved with a `.sas` extension) which might read in or perhaps modify a saved dataset. Then a `proc` is written to perform a desired analysis. The output is examined, and based on the results, the `data` step is modified to generate new variables, the `proc` is edited to choose new options, new `procs` are written, or some subset of these steps is repeated. At the end of the session, the dataset might be saved in the native SAS format, the commands saved in text format, and the results saved electronically or printed.

1.6 Useful SAS background

1.6.1 Dataset options

In addition to `data` steps for manipulating data, SAS allows on-the-fly modification of datasets. This approach, while less than ideal for documentation, can be a useful way to reduce code length: rather than create a new dataset with a subset of observations, or with a renamed variable, this can be done simultaneously with specifying the dataset to be used in a procedure. The syntax for these commands, called "dataset options" in SAS documentation, is to list them in parentheses after naming the dataset. So, for example, to temporarily exclude extraneous variables in a dataset from an analysis dataset, the following code could be used to save time if the dataset were large.

```
proc ttest data=test2 (keep=x y);
  class x;
  var y;
run;
```

Another useful dataset option limits the number of observations used in the procedure.

```
proc ttest data=test2 (obs=60);
  class x;
  var y;
run;
```

A full list of dataset options can be found in the online documentation: Contents; SAS Products; Base SAS; SAS 9.2 Language Reference: Dictionary; Dictionary of Language Elements; SAS Data Set Options.

1.6.2 Repeating commands for subgroups

A common problem in data analysis involves repeating some process for strata defined by a categorical variable. For example, a model might be needed for males and females separately, or for several different age groups. SAS provides an easy way to do this via the `sort` procedure and the `by` statement. Here we demonstrate using variables from the HELP dataset, assuming it has been read in using one of the methods described in Section 2.1 and demonstrated at the outset of each example section.

```
proc sort data=ds;
  by female;
run;

proc glm data=ds;
  by female;
  model mcs = pcs;
run;
```

The `proc glm` code will generate regression output for each value of `female`. Many procedures support a `by` statement in this fashion. If the data have not been `sorted` previously, an error is likely.

1.6.3 Subsetting

It is often convenient to restrict the membership in a dataset or run analyses on a subset of observations. There are three main ways we do this in SAS. One is through the use of a subsetting `if` statement in a data step. The syntax for this is simply

```
data ...;
set ...;
  if condition;
run;
```

where condition is a logical statement such as `x eq 2` (see 2.11.2 for a discussion of logical operators). This includes only observations for which the condition is true, because when an `if` statement (2.11.2) does not include a `then`, the implied `then` clause is interpreted as "then output this line to the dataset; otherwise do not output it."

A second approach is a `where` statement. This can be used in a `data` step or in a procedure, and has a similar syntax.

```
proc ... data=ds;
  where condition;
  ...
run;
```

Finally, there is also a `where` dataset option which can be used in a `data` step or a procedure; the syntax here is slightly different.

```
proc ... data=ds (where=(condition));
  ...
run;
```

The differences between the `where` statement and the `where` dataset option are subtle and beyond our scope here. However, it is generally computationally cheaper to use a `where` approach than a subsetting `if`.

1.6.4 Formats and informats

SAS provides special tools for displaying variables or reading them in when they have complicated or unusual constructions in raw text. A good example for this is dates, for which June 27, 2009 might be written as, for example, `6-27-12, 27-6-12, 06/27/2012`, and so on. SAS stores dates as the integer number of days since December 31, 1959. To convert one of the aforementioned expressions to the desired storage value, `17710`, you use an *informat* to describe the way the data is written. For example, if the data were stored as the above expressions, you would use the informats `mmddyy8.`, `ddmmyy8.`, and `mmddyy10.` respectively to read them correctly as `17710`. An example of reading in dates is shown in Section 2.1.2. More information on informats can be found in the online documentation: Contents; SAS Products; Base SAS; SAS 9.2 Language Reference: Dictionary; Informats.

In contrast, displaying data in styles other than that in which it is stored is done using the `informat`'s inverse, the `format`. The format for display can be specified within a `proc`. For example, if we plan a time series plot of `x*time` and want the x-axis labeled in quarters (i.e., `2010Q3`), we could use the following code, where the `time` variable is the integer-valued date. Information on formats can be found in the online documentation: Contents; SAS Products; Base SAS; SAS 9.2 Language Reference: Dictionary; Formats.

```
proc gplot data=ds;
  plot x*time;
  format time yyq6.;
run;
```

Another example is deciding how many decimal digits to display. For example, if you want to display 2 decimal places for variable p and 3 for variable x, you could use the following code.

```
proc print data=ds;
  var p x;
  format p 4.2 x 5.3;
run;
```

This topic is also discussed in Section 2.2.4.

1.7 Accessing and controlling SAS output: The Output Delivery System

SAS does not generally provide access to most of the internal objects used in calculating results. Instead, it provides specific access to many objects of interest through various procedure statements. The ways to find these objects is somewhat idiosyncratic, and we have tried to highlight the most commonly needed objects in the text.

A much more general way to access and control output within SAS is through the Output Delivery System, also called (redundantly, as in "ATM machine") the ODS system. This is a very powerful and flexible system for accessing procedure results and controlling printed output. We use the ODS system mainly for two tasks: (1) to save procedure output into explicitly named datasets and (2) to suppress some printed output from procedures which generate lengthy output. In addition, we discuss using the ODS system to save output in useful file formats such as portable document format (PDF), hypertext markup language (HTML), or rich text format (RTF). Finally, we discuss ODS graphics, which add graphics to procedures' text output. We note that ODS has other uses beyond what we show, and encourage readers to spend time familiarizing themselves with it. The best resource for this is the online help: Contents; SAS Products; Base SAS; SAS 9.2 Output Delivery System User's Guide.

1.7.1 Saving output as datasets and controlling output

Using ODS to save output or control the printed results involves two steps; first, finding out the name by which the ODS system refers to the output, and second, requesting that the dataset be saved as an ordinary SAS dataset or including or excluding it as output. The names used by the ODS system can be most easily found by running an `ods trace on / listing` statement (later reversed using an `ods trace off` statement). The ods `outputname` thus identified can be saved using an `ods output outputname=newname` statement. A piece of output can be excluded using an `ods exclude outputname1 outputname2 ... outputnamek` statement or only desired pieces printed using the `ods select outputname1 outputname2 ... outputnamek` statement. These statements are each included before the procedure code which generates the output concerned. The `exclude` and `select` statements can be reversed using an `ods exclude none` or `ods select all` statement.

For example, to save the result of the t-test performed by `proc ttest` (3.4.1), the following code would be used. First, we generate some data for the test.

```
data test2;
  do i = 1 to 100;
    if i lt 51 then x=1;
        else x=0;
    y = normal(0) + x;
    output;
  end;
run;
```

Then, run the t-test, including the `ods trace on / listing` statement to learn the names used by the `ODS` system.

```
ods trace on / listing;
proc ttest data=test2;
  class x;
  var y;
run;
ods trace off;
```

This would result in the following output.

```
Variable:  y

Output Added:
-------------
Name:       Statistics
Label:      Statistics
Template:   Stat.TTest.Statistics
Path:       Ttest.y.Statistics
-------------

x                N        Mean      Std Dev      Std Err

0               50      0.0193       1.1552       0.1634
1               50      1.0903       0.9686       0.1370
Diff (1-2)             -1.0710       1.0660       0.2132

x              Minimum     Maximum

0              -2.1863      2.5731
1              -0.8381      3.5587
Diff (1-2)
```

```
Variable:  y

Output Added:
-------------
Name:       ConfLimits
Label:      Confidence Limits
Template:   Stat.TTest.ConfLimits
Path:       Ttest.y.ConfLimits
-------------

x            Method              Mean      95% CL Mean        Std Dev

0                              0.0193    -0.3090    0.3476     1.1552
1                              1.0903     0.8150    1.3656     0.9686
Diff (1-2)   Pooled           -1.0710    -1.4941   -0.6479     1.0660
Diff (1-2)   Satterthwaite    -1.0710    -1.4942   -0.6477

x            Method           95% CL Std Dev

0                              0.9650    1.4395
1                              0.8091    1.2070
Diff (1-2)   Pooled            0.9354    1.2393
Diff (1-2)   Satterthwaite
```

```
Variable:  y

Output Added:
-------------
Name:       TTests
Label:      T-Tests
Template:   Stat.TTest.TTests
Path:       Ttest.y.TTests
-------------

Method           Variances        DF    t Value    Pr > |t|

Pooled           Equal            98     -5.02       <.0001
Satterthwaite    Unequal      95.107     -5.02       <.0001
```

```
Variable:  y

Output Added:
-------------
Name:       Equality
Label:      Equality of Variances
Template:   Stat.TTest.Equality
Path:       Ttest.y.Equality
-------------

              Equality of Variances

Method      Num DF    Den DF    F Value    Pr > F

Folded F        49        49       1.42    0.2209
```

Note that failing to issue the `ods trace off` command will result in continued annotation of every piece of output. Similarly, when using the `ods exclude` and `ods select` statements, it is good practice to conclude each procedure with an `ods select all` or `ods exclude none` statement so that later output will be printed.

The previous output shows that the t-test itself (including the tests assuming equal and unequal variances) appears in output which the `ODS` system calls `ttests`, so the following code demonstrates how the test can be saved into a new dataset. Here we assign the new dataset the name `newttest`.

```
ods output ttests=newttest;
proc ttest data=test2;
  class x;
  var y;
run;

proc print data=newttest;
run;
```

The `proc print` code results in the following output.

```
Obs  Variable  Method          Variances   tValue      DF   Probt

  1     y      Pooled           Equal       -5.02      98  <.0001
  2     y      Satterthwaite   Unequal      -5.02  95.107  <.0001
```

To run the t-test and print only these results, the following code would be used.

```
options ls=64;
ods select ttests;
proc ttest data=test2;
  class x;
  var y;
run;
ods select all;

Variable:  y

Method           Variances       DF    t Value    Pr > |t|

Pooled           Equal           98      -5.02      <.0001
Satterthwaite    Unequal     95.107      -5.02      <.0001
```

This application is especially useful when running simulations, as it allows the results of procedures to be easily stored for later analysis.

The foregoing barely scratches the surface of what is possible using `ODS`. For further information, refer to the online help: Contents; SAS Products; Base SAS; SAS 9.2 Output Delivery System User's Guide.

1.7.2 Output file types and ODS destinations

The other main use of `ODS` is to generate output in a variety of file types. By default, SAS output is printed in the `output` window in the internal GUI. When run in batch mode, or when saving the contents of the output window using the GUI, this output is saved as a plain text file with a `.lst` extension. The `ODS` system provides a way to save SAS output in a more attractive form. As discussed in Section 6.4, procedure output and graphics can be saved to named output files by using commands of the following form.

```
ods destinationname file="filename.ext";
```

The valid `destinationnames` include `pdf`, `rtf`, `latex`, and others. SAS refers to these file types as "destinations." It is possible to have multiple destinations open at the same time. For destinations other than `listing` (the output window), the `destination` must be closed before the results can be seen. This is done using the `ods destinationname close` statement. Note that the default `listing` destination can also be closed; if there are no output destinations open, no results can be seen.

1.7.3 ODS graphics

The `ODS` system also allows users to incorporate text and graphical output from a procedure in an integrated document. This is done by "turning on" `ODS` graphics using an `ods graphics on` statement (as demonstrated in Section 5.6.8), and then accepting default graphics or requesting particular plots using a `plots=plotnames` option to the procedure statement, where the valid plot names vary by procedure.

Special note for UNIX users: To generate ODS Graphics output in UNIX batch jobs, you must set the DISPLAY system option before creating the output. To set the display, enter the following command in the shell.

```
export DISPLAY=<ip_address>:0 (Korn shell)

DISPLAY=<ip_address>:0
export DISPLAY          (Bourne shell)

setenv DISPLAY=<ip_address>:0 (C shell)
```

In the above, `ip_address` is the fully qualified domain name or IP address, or the name of a display. Usually, the IP address of the UNIX system where SAS is running would be used. If you do not set the DISPLAY variable, then you get an error message in the SAS log. Additional information for UNIX users can be found in the online help: Contents; Using SAS Software in Your Operating Environment; SAS 9.2 Companion for UNIX Environments.

1.8 The SAS Macro Facility: Writing functions and passing values

1.8.1 Writing functions

SAS does not provide a simple means for constructing functions which can be integrated with other code. However, it does provide a text-replacement capacity called the SAS Macro Language which can simplify and shorten code. The language also includes looping capabilities. We demonstrate here a simple macro to change the predictor in a simple linear regression example.

```
%macro reg1 (pred=);
  proc reg data=ds;
    model y = &pred;
  run;
%mend reg1;
```

In this example, we define the new macro by name (`reg1`) and define a single parameter which will be passed in the macro call; this will be referred to as `pred` within the macro. To replace `pred` with the submitted value, we use `&pred`. Thus the macro will run `proc reg` (Section 4.1.1) with whatever text is submitted as the predictor of the outcome `y`. This macro would be called as follows.

```
%reg1(pred=x1);
```

When the `%macro` statements and the `%reg1` statement are run, SAS sees the following.

```
proc reg data=ds;
  model y = x1;
run;
```

If four separate regressions were required, they could then be run in four statements.

```
%reg1(pred=x1);
%reg1(pred=x2);
%reg1(pred=x3);
%reg1(pred=x4);
```

As with the Output Delivery System, SAS macros are a much broader topic than can be fully explored here. For a full introduction to its uses and capabilities, see the online help: Contents; SAS Products; Base SAS; SAS 9.2 Macro Language: Reference.

1.8.2 SAS macro variables

SAS also includes what are known as *macro variables*. Unlike SAS macros, macro variables are values that exist during SAS runs and are not stored within datasets. In general, a macro variable is defined with a `%let` statement.

```
%let macrovar=chars;
```

Note that the `%let` statement need not appear within a `data` step; it is a global statement. The value is stored as a string of characters, and can be referred to as `¯ovar`. For example:

```
data ds;
  newvar=&macrovar;
run;
```

or

```
title "This is the &macrovar";
```

In the above example, the double quotes in the `title` statement allow the text within to be processed for the presence of macro variables. Enclosing the title text in single quotes will result in `¯ovar` appearing in the title, while the code above will replace `¯ovar` with the value of the `macrovar` macro variable.

While this basic application of macro variables is occasionally useful in coding, a more powerful application is to generate the macro variables within a SAS `data` step. This can be done using a `call symput` function as shown in 3.6.4.

```
data _null_;
  ...
  call symput('macrovar', x);
run;
```

This makes a new macro variable named `macrovar` which has the value of the variable `x`. It can be accessed as `¯ovar`. The `_null_` dataset is a special SAS dataset which is not saved. It is efficient to use it when there is no need for a stored dataset.

1.9 Interfaces: Code and menus, data exploration, and data analysis

We find the SAS windows system to be a comfortable work environment and an aid to productivity. However, SAS can be easily run in batch mode. To use SAS this way, compose code in the text editor of your choice. Save the file (a `.sas` extension would be appropriate), then find it in the operating system. In Windows, a right-click on the file will bring up a list of potential actions, one of which is "Batch Submit with SAS 9.2." If this option is selected, SAS will run the file without opening the GUI. The output will be saved in the same directory with the same name but with a `.lst` extension; the log will be saved in the same directory with the same name but with a `.log` extension. Both output files are plain text files.

We prefer the code-based approach featured in this book, which allows the fine control often necessary in data analysis as well as the easy replicability and succinct clarity which code provides, but allows for more speed than a batch-mode approach. However, some people may prefer a menu-based interface to analytic tools, especially for exploring data. SAS provides several tools for such an approach. The SAS/Analyst application can be started from the menu system, under the `Solutions` tab: `Analysis` and `Analyst`. For information, see `http://support.sas.com/rnd/app/da/analyst/overview.html` or the online help: Contents; SAS Products; SAS/STAT; Analyst Application User's Guide. Another option is SAS/INSIGHT, which resembles SAS/Analyst. SAS/INSIGHT can be started from the menu system, under the `Solutions` tab under `Analysis` and `Interactive Data Analysis`. It can also be accessed via code using `proc insight`. For information, see `http://support.sas.com/documentation/onlinedoc/insight/index.html` or the online help: Contents; SAS Products; SAS/INSIGHT User's Guide. A third similar approach is included in SAS Stat Studio. This will be called SAS/IML Studio as on SAS 9.2, and will be the only one of these three tools available in future releases of SAS. In SAS 9.2, SAS/IML Studio requires a separate download and installation. Another product is SAS/LAB; see `http://www.sas.com/products/lab` or the online help: Contents; SAS Products; SAS/LAB Software for more information.

1.10 Miscellanea

Official documentation provided by SAS refers to, for example `PROC GLM`. However, SAS is not case sensitive, with a few exceptions. In this text we use lower case throughout. We find lower case easier to read, and prefer the ease of typing, both for coding and book composition, in lower case.

Since statements are separated by semicolons, multiple statements may appear on one line and statements may span lines. We usually type one statement per line in the text (and in practice), however. This prevents statements being overlooked among others appearing in the same line. In addition, we indent statements within a data step or proc, to clarify the grouping of related commands.

SAS includes both `run` and `quit` statements. The `run` statement tells SAS to act on the code submitted since the most recent `run` statement (or since startup, if there has been no `run` statement submitted thus far). Some procedures allow multiple steps within the procedure without having to end it; key procedures which allow this are `proc gplot` and `proc reg`. This might be useful for model fitting and diagnostics with particularly large datasets in `proc reg`. In general, we find it a nuisance in graphics procedures, because the graphics are sometimes not actually drawn until the `quit` statement is entered. In the examples, we use the `run` statement in general and the `quit` statement when necessary, without further comment.

Chapter 2

Data management

This chapter reviews basic data management, beginning with accessing external datasets, such as those stored in spreadsheets, ASCII files, or foreign formats. Important tasks such as creating datasets and manipulating variables are discussed in detail. In addition, key mathematical, statistical, and probability functions are introduced.

2.1 Input

SAS native datasets are rectangular files with data stored in a special format. They have the form `filename.sas7bdat` or something similar, depending on version. In the following, we assume that files are stored in directories and that the locations of the directories in the operating system can be labeled using Windows syntax (though SAS allows UNIX/Linux/Mac OS X-style forward slash as a directory delimiter on Windows). Other operating systems will use local idioms in describing locations.

Sections 2.1.2 through 2.1.7 show code-based methods for importing data into SAS when they are stored in other formats. This approach is especially important in the common case where new versions or similar files must be repeatedly accessed. However, it may also be helpful or effective to use the *Import Data* wizard. This is a point-and-click dialog that is started from the *File* menu. It is simple to use and robust. As of SAS 9.2, it provides access to files in the formats native to Microsoft Excel, Microsoft Access, delimited text, dBase, JMP, Lotus 1-2-3, SPSS, Stata, and Paradox. The wizard works by using the dialog to write `proc import` code, just like the code we use in Sections 2.1.2, 2.1.4, and 2.1.5. One particularly useful feature of the wizard is that it gives the option of saving the `proc import` code thus generated in a file. So the wizard can be used once to generate code that can be reapplied later without having to run through the steps required by the wizard.

2.1.1 Native dataset

Example: See 5.6

```
libname libref "dir_location";
data ds;
  set libref.sasfilename; /* Note: no file extension */
  ...
run;
```

or

```
data ds;
  set "dir_location\sasfilename.sas7bdat";  /* Windows only */
  set "dir_location/sasfilename.sas7bdat";
  /* works on all OS including Windows */
  ...
run;
```

The file `sasfilename.sas7bdat` is created by using a `libref` in a `data` statement; see 2.2.1. The set statement has several options, including `end=varname`, which creates a variable `varname` which is 1 in the last observation and 0 otherwise, and `nobs=varname` which contains the total number of observations in the dataset. Both can be very useful in coding, and in either case, the variables are not saved in the output dataset.

For example, consider the following code.

```
data ds2;
  set ds end=last nobs=numds;
  if numds ge 10 and last;
run;
```

The output dataset `ds2` will contain the last observation of the input dataset `ds` if there are at least 10 observations in `ds`. Otherwise `ds2` will have 0 observations.

2.1.2 Fixed format text files

See also 2.1.3 (read more complex fixed files) and 7.4 (read variable format files)

```
data ds;
  infile 'C:\file_location\filename.ext';  /* Windows only */
  input varname1 ... varnamek;
run;
```

or

```
filename filehandle 'file_location/filename.ext';

proc import datafile=filehandle
  out=ds dbms=dlm;
  getnames=yes;
run;
```

The `infile` approach allows the user to limit the number of rows read from the data file using the `obs` option. Character variables are noted with a trailing '$', e.g., use a statement such as `input varname1 varname2 $ varname3` if the second position contains a character variable (see 2.1.3 for examples). The `input` statement allows many options and can be used to read files with variable format (7.4.1).

In `proc import`, the `getnames=yes` statement is used if the first row of the input file contains variable names (the variable types are detected from the data). If the first row does not contain variable names then the `getnames=no` option should be specified. The `guessingrows` option (not shown) will base the variable formats on other than the default 20 rows. The `proc import` statement will accept an explicit file location rather than a file associated by the `filename` statement as in Section 5.6.

Note that in Windows installations, SAS accepts either slashes or backslashes to denote directory structures. For Linux, only forward slashes are allowed. Behavior in other operating systems may vary.

In addition to these methods, files can be read by selecting the `Import Data` option on the `file` menu in the GUI.

2.1.3 Reading more complex text files

See also 2.1.2 (read fixed files) and 7.4 (read variable format files)

Text data files often contain data in special formats. One common example is date variables. Special values can be read in using informats (1.6.4). As an example below we consider the following data.

```
1 AGKE 08/03/1999 $10.49
2 SBKE 12/18/2002 $11.00
3 SEKK 10/23/1995 $5.00
```

```
data ds;
  infile 'C:\file_location\filename.dat';
  input id initials $ datevar mmddyy10. cost dollar7.4;
run;
```

The SAS informats (Section 1.6.4) denoted by the `mmddyy10.` and `dollar7.4` inform the `input` statement that the third and fourth variables have special values and should not be treated as numbers or letters, but read and interpreted according to the specified rules. In the case of `datevar`, SAS reads the date appropriately and stores a SAS date value (Section 1.6.4). For `cost`, SAS ignores the '$' in the data and would also ignore commas, if they were present. The `input` statement allows many options for additional data formats and can be used to read files with variable format (7.4.1).

Other common features of text data files include very long lines and missing data. These are addressed through the `infile` or `filename` statements. Missing data may require the `missover` option to the `infile` statement as well as listing the columns in which variables appear in the dataset in the `input` statement. Long lines (many columns in the data file) may require the `lrecl` option to the `infile` or `filename` statement. For a thorough discussion, see the online help: Contents; SAS Products; Base SAS; SAS 9.2 Language Reference: Concepts; DATA Step Concepts; Reading Raw Data; Reading Raw Data with the INPUT statement.

2.1.4 Comma-separated value (CSV) files

Example: See 2.13.1

```
data ds;
  infile 'dir_location/filename.csv' delimiter=',';
  input varname1 ... varnamek;
run;
```

or

```
proc import datafile='dir_location\full_filename'
  out=ds dbms=dlm;
  delimiter=',';
  getnames=yes;
run;
```

Character variables are noted with a trailing '$', e.g., use a statement such as `input varname1 varname2 $ varname3` if the second column contains characters. The `proc import` syntax allows for the first row of the input file to contain variable names, with variable types detected from the data. If the first row does not contain variable names then use `getnames=no`.

In addition to these methods, files can be read by selecting the `Import Data` option on the `file` menu in the GUI.

2.1.5 Reading datasets in other formats

Example: See 4.7.1

```
libname ref spss 'filename.sav'; /* SPSS */
libname ref bmdp 'filename.dat'; /* BMDP */
libname ref v6 'filename.ssd01; /* SAS vers. 6 */
libname ref xport 'filename.xpt'; /* SAS export */
libname ref xml 'filename.xml'; /* XML */

data ds;
  set ref.filename;
run;
```

or

```
proc import datafile="filename.ext' out=ds
  dbms=excel; /* Excel */
run;

  ... dbms=access; ... /* Access */
  ... dbms=dta; ... /* Stata */
```

The `libname` statements above refer to files, rather than directories. The extensions shown above are those typically used for these file types, but in any event the full name of the file, including the extension, is needed in the `libname` statement. In contrast, only the file name (without the extension) is used in the `set` statement. The data type options specified above in the `libname` statement and `dbms` option are available in Windows. To see what is available under other operating systems, check in the online help: Contents, Using SAS in Your Operating Environment, SAS 9.2 Companion for <your OS>, Features of the SAS language for <your OS>, Statements under <your OS>, Libname statement.

In addition to these methods, files can be read by selecting the `Import Data` option on the `file` menu in the GUI.

2.1.6 URL

Example: See 3.6.1

```
filename urlhandle url 'http://www.math.smith.edu/sas/testdata';

filename urlhandle url 'http://www.math.smith.edu/sas/testdata'
  user='your_username' pass='your_password';

proc import datafile=urlhandle out=ds dbms=dlm;
run;
```

The latter `filename` statement is needed only if the URL requires a username and password. The `urlhandle` used in a `filename` statement can be no longer than 8 characters. A `urlhandle` can be used in an `import` procedure as shown, or with an `infile` statement in a `data` step (see 7.4). The `import` procedure supports many file types through the `dbms` option; `dbms=dlm` without the `delimiter` option (Section 2.1.4) is for space-delimited files.

2.1.7 XML (extensible markup language)

A sample (flat) XML form of the HELP dataset can be found at `http://www.math.smith.edu/sas/datasets/help.xml`. The first 10 lines of the file consist of:

```
<?xml version="1.0" encoding="iso-8859-1" ?>
<TABLE>
   <HELP>
      <id> 1 </id>
      <e2b1 Missing="." />
      <g1b1> 0 </g1b1>
      <i11 Missing="." />
      <pcs1> 54.2258263 </pcs1>
      <mcs1> 52.2347984 </mcs1>
      <cesd1> 7 </cesd1>
```

Here we consider reading simple files of this form. While support is available for reading more complex types of XML files, these typically require considerable additional sophistication.

```
libname ref xml 'dir_location/filename.xml';

data ds;
  set ref.filename_without_extension;
run;
```

The `libname` statement above refers to a file name, rather than a directory name. The “xml” extension is typically used for this file type, but in any event the full name of the file, including the extension, is needed.

2.1.8 Data entry

See also 2.2.5 (spreadsheet data access)

```
data ds;
  input x1 x2;
  cards;
1 2
1 3
1.4 2
123 4.5
;
run;
```

The above code demonstrates reading data into a SAS dataset within a SAS program. The semicolon following the data terminates the `data` step, meaning that a `run` statement is not actually required. The `input` statement used above employs the syntax discussed in 2.1.2. In addition to this option for entering data within SAS, there is a GUI-based data entry/editing tool called the Table Editor. It can be accessed using the mouse through the Tools menu, or by using the `viewtable` command on the SAS command line.

2.2 Output

2.2.1 Save a native dataset

Example: See 2.13.1

```
libname libref "dir_location";

data libref.sasfilename;
  set ds;
run;
```

A SAS dataset can be read back into SAS using a `set` statement with a `libref`, see 2.1.1.

2.2.2 Creating files for use by other packages

See also 2.2.8 (write XML) *Example:* See 2.13.1

```
libname ref spss 'filename.sav'; /* SPSS */
libname ref bmdp 'filename.dat'; /* BMDP */
libname ref v6 'filename.ssd01'; /* SAS version 6 */
libname ref xport 'filename.xpt'; /* SAS export */
libname ref xml 'filename.xml'; /* XML */

data ref.filename_without_extension;
  set ds;
run;
```

or

```
proc export data=ds outfile='file_location_and_name'
  dbms=csv; /* comma-separated values */

  ...dbms=dbf; /* dbase 5,IV,III */
  ...dbms=excel; /* excel */
  ...dbms=tab; /* tab-separated values */
  ...dmbs=access; /* Access table */
  ...dbms=dlm; /* arbitrary delimiter; default is space,
                  others with delimiter=char statement */
```

The `libname` statements above refer to file names, rather than directory names. The extensions shown above are those conventionally used but the option specification determines the file type that is created.

2.2.3 Creating datasets in text format

```
proc export data=ds outfile='file_location_and_name'
  dbms=csv; /* comma-separated values */
run;

  ...dbms=tab; /* tab-separated values */
  ...dbms=dlm; /* arbitrary delimiter */
```

For `dbms=dlm`, the default delimiter is a space. If another delimiter is needed, add a separate `delimiter='x'` statement, where `x` is the delimiting character.

2.2.4 Displaying data

Example: See 4.7.3

See also 2.3.2 (values of variables in a dataset) and 2.2.5 (spreadsheet access to data)

```
title1 'Display of variables';
footnote1 'A footnote';
proc print data=ds;
  var x1 x3 xk x2;
  format x3 dollar10.2;
run;
```

For `proc print` the `var` statement specifies variables to be displayed. The format statement, as demonstrated, can alter the appearance of the data; here x_3 is displayed as a dollar amount with 10 total digits, two of them to the right of the decimal. The keyword `_numeric_` can replace the variable name and will cause all of the numerical variables to be displayed in the same format. See Section 1.6.4 for further discussion.

See Sections 1.6.3, 1.6.2, and 1.6.1 for ways to limit which observations are displayed. The `var` statement, as demonstrated, can cause the variables to be displayed in the desired order. The `title` and `footnote` statements and related statements `title1`, `footnote2`, etc., allow headers and footers to be added to each output page. Specifying the command with no argument will remove the title or footnote from subsequent output.

SAS also provides `proc report` and `proc tabulate` to create more customized output.

2.2.5 Spreadsheet data display and editing

See also 2.2.4 (displaying data) and 2.3.2 (values of variables)

As a flexible and easy alternative to `proc print` (2.3.2), SAS includes a spreadsheet-like data viewing and entry tool. This is called *viewtable*. It cannot be accessed by submitting code in the program window but by clicking on *Table Editor* in the *Tools* menu, or by typing `viewtable` in the command window, as shown in Figure 2.1. The tool is started by pressing the enter key or clicking the check mark shown to the left of the command window.

Once started, a blank table is shown, as in Figure 2.2. The blank table can be used for data entry and saved (using the anachronistic floppy disc icon), after which it is available for analysis.

To open an existing SAS dataset, click the opening folder icon. This pops up a list of available SAS libraries, including the `WORK` library in which SAS stores all datasets by default, any library previously specified in a `libname` statement (2.1.1), and several libraries installed with SAS. After finding and opening the desired dataset, it is displayed in familiar and navigable spreadsheet form, and shown in Figure 2.3. In this case we display the HELP dataset (A). Note that

Figure 2.1: Starting *viewtable* from the command window.

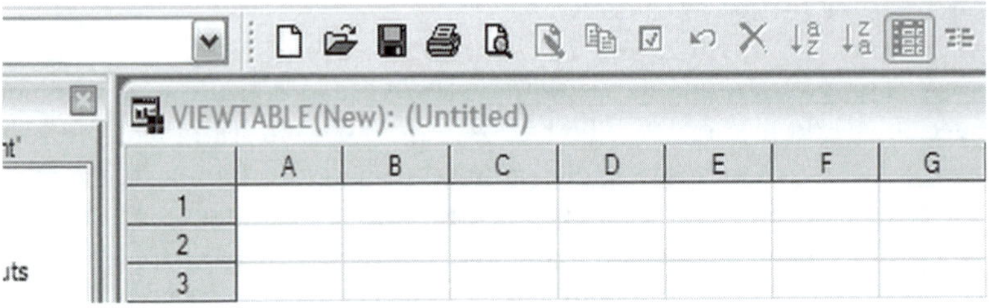

Figure 2.2: After starting *viewtable*.

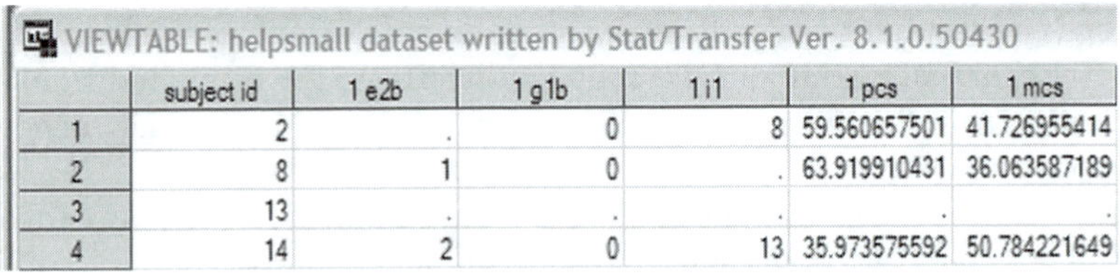

Figure 2.3: Displaying existing data with *viewtable*.

existing data can be edited using *viewtable*. For real applications this is rarely wise, as it is difficult to replicate and record the change.

2.2.6 Number of digits to display

Example: See 2.13.1

SAS lacks an option to control how many significant digits are displayed in procedure output, in general (an exception is `proc means`). For reporting purposes, one should save the output as a dataset using `ODS`, then use the `format` statement (2.2.4 and 1.6.4) with `proc print` to display the desired precision as demonstrated in Section 4.7.3.

2.2.7 Creating HTML formatted output

```
ods html file="filename.html";
...
ods html close;
```

Any output generated between an `ods html` statement and an `ods hmtl close` statement will be included in an HTML (hypertext markup language) file (1.7.2). By default this will be displayed in an internal SAS window; the optional `file` option shown above will cause the output to be saved as a file.

2.2.8 Creating XML datasets and output

```
libname ref xml 'dir_location/filename.xml';

data ref.filename_without_extension;
  set ds;
run;
```

or

```
ods docbook file='filename.xml';
...
ods close;
```

The `libname` statement can be used to write a SAS dataset to an XML-formatted file. It refers to a file name, rather than a directory name. The file extension `xml` is conventionally used but the `xml` specification, rather than the file extension, determines the file type that is created.

The `ods docbook` statement, in contrast, can be used to generate an XML file displaying procedure output; the file is formatted according to the OASIS DocBook DTD (document type definition).

2.3 Structure and meta-data

2.3.1 Names of variables and their types

Example: See 2.13.1

```
proc contents data=ds;
run;
```

2.3.2 Values of variables in a dataset

Example: See 2.13.2

See also 2.2.5 (spreadsheet access to data) and 2.2.4 (displaying data)

```
proc print data=ds (obs=nrows);
  var x1 ... xk;
run;
```

The integer `nrows` for the `obs=nrows` option specifies how many rows to display, while the `var` statement selects variables to be displayed (1.6.1). Omitting the `obs=nrows` option or `var` statement will cause all rows and all variables in the dataset to be displayed, respectively.

2.3.3 Rename variables in a dataset

```
data ds2;
  set ds (rename = (old1=new1 old2=new2 ...));
  ...
run;
```

or

```
data ds;
  ...
rename old=new;
run;
```

2.3.4 Add comment to a dataset or variable

Example: See 2.13.1

To help facilitate proper documentation of datasets, it can be useful to provide some annotation or description.

```
data ds (label="This is a comment about the dataset");
...
run;
```

The label can be viewed using `proc contents` (2.3.1) and retrieved as data using `ODS` (see 1.7).

2.4 Derived variables and data manipulation

This section describes the creation of new variables as a function of existing variables in a dataset.

2.4.1 Create string variables from numeric variables

```
data ...;
  stringx = input(numericx, $char.);
run;
```

Applying any string function to a numeric variable will force (coerce) it to be treated as a character variable. As an example, concatenating (see 2.4.5) two numeric variables (i.e., `v3 = v1||v2`) will result in a string. See 1.6.4 for a discussion of informats, which apply variable types when reading in data.

2.4.2 Create numeric variables from string variables

```
data ...;
  numericx = input(stringx, integer.decimal);
run;
```

In the argument to the `input` function, `integer` is the number of characters in the string, while `decimal` is an optional specification of how many characters appear after the decimal.

Applying any numeric function to a variable will force it to be treated as numeric. For example: a `numericx = stringx * 1.0` statement will also make `numericx` a numeric variable.

See also 1.6.4 for a discussion of informats, which apply variable types when reading in data.

2.4.3 Extract characters from string variables

See also 2.4.8 (replace strings within strings)

```
data ...;
  get2through4 = substr(x, 2, 3);
run;
```

The syntax functions as follows: name of the variable, start character place, how many characters to extract. The last parameter is optional. When omitted, all characters after the location specified in the second space will be extracted.

2.4.4 Length of string variables

```
data ...;
  len = length(stringx);
run;
```

In this example, `len` is a variable containing the number of characters in `stringx` for each observation in the dataset, excluding trailing blanks. Trailing blanks can be included through use of the `lengthc` function.

2.4.5 Concatenate string variables

```
data ...;
  newcharvar = x1 || " VAR2 " x2;
run;
```

The above SAS code creates a character variable `newcharvar` containing the character variable X_1 (which may be coerced from a numeric variable) followed by the string `" VAR2 "` then the character variable X_2. By default, no spaces are added.

2.4.6 Find strings within string variables

```
data ...;
  /* where is the first occurrence of "pat"? */
  match = find(stringx, "pat");
  /* where is the first occurrence of "pat" after startpos? */
  matchafter = find(stringx, "pat", startpos);
  /* how many times does "pat" appear? */
  howmany = count(stringx, "pat");
run;
```

Without the option `startpos`, `find` returns the start character for the first appearance of `pat`. If `startpos` is positive, the search starts at `startpos`, if it is negative, the search is to the left, starting at `startpos`. If `pat` is not found or `startpos=0`, then `match=0`.

2.4.7 Remove characters from string variables

See also 2.4.9 (remove spaces around strings)

```
data ...;
  nospaces = compress(stringx);
  noletteras = compress(stringx, 'a');
run;
```

By default, spaces are removed. If there is a second argument, all characters listed there will be removed.

2.4.8 Replace characters within string variables

See also 2.4.3 (extract characters from strings)

```
data ...;
  substr(stringx,1,3) = "Boy";
run;
```

The first three characters of the variable `stringx` are now `Boy`. This application of the `substr` function uses the same syntax as when used to extract characters (2.4.3).

2.4.9 Remove spaces around string variables

```
data ...;
  noleadortrail = strip(stringx);
run;
```

The `trimn(stringx)` function will remove only the trailing blanks.

2.4.10 Upper to lower case

```
data ...;
  lowercasex = lowcase(x);
run;
```

or

```
data ...;
  lowercasex = translate(x, "abcdefghijklmnopqrstuvwxzy",
  "ABCDEFGHIJKLMNOPQRSTUVWXYZ");
run;
```

The upcase function makes all characters upper case. Arbitrary translations from sets of characters can be made using the translate function.

2.4.11 Create categorical variables from continuous variables

Example: See 2.13.3 and 5.6.6

```
data ...;
  if x ne . then newcat = (x ge minval) + (x ge cutpoint1) +
  ... + (x ge cutpointn);
run;
```

Each expression within parentheses is a logical test returning 1 if the expression is true, 0 otherwise. If the initial condition is omitted then a missing value for x will return the value of 0 for newcat. More information about missing value coding can be found in Section 2.4.16 (see 2.11.2 for more about conditional execution).

2.4.12 Recode a categorical variable

A categorical variable may need to be recoded to have fewer levels.

```
data ...;
  newcat = (oldcat in (val1, val2, ..., valn)) +
  (oldcat in (val1, val3)) + ...;
run;
```

The in function can also accept quoted strings as input. It returns a value of 1 if any of the listed values is equal to the tested value.

2.4.13 Create a categorical variable using logic

Example: See 2.13.3

Here we create a trichotomous variable newvar which takes on a missing value if the continuous non-negative variable oldvar is less than 0, 0 if the continuous variable is 0, value 1 for subjects in group A with values greater than 0 but less than 50 and for subjects in group B with values greater than 0 but less than 60, or value 2 with values above those thresholds.

More information about missing value coding can be found in Section 2.4.16.

```
data ...;
  if oldvar le 0 then newvar=.;
  else if oldvar eq 0 then newvar=0;
  else if ((oldvar lt 50 and group eq "A") or
    (oldvar lt 60 and group eq "B"))
  then newvar=1;
  else newvar=2;
run;
```

2.4.14 Formatting values of variables

Example: See 4.7.3

Sometimes it is useful to display category names that are more descriptive than variable names. In general, we do not recommend using this feature (except potentially for graphical output), as it tends to complicate communication between data analysts and other readers of output (see also labeling variables, 2.4.15). In this example, character labels are associated with a numeric variable (0=Control, 1=Low Dose and 2=High Dose).

```
proc format;
  value dosegroup 0 = 'Control' 1 = 'Low Dose' 2 = 'High Dose';
run;
```

Many procedures accept a `format x dosegroup.` statement (note trailing '.'); this syntax will accept formats designed by the user with the `proc format` statement, as well as built-in formats (see 2.2.4). Categorizations of a variable can also be imposed using `proc format`, but this can be cumbersome. In all cases, a new variable should be created as described in 2.4.11 or 2.4.12.

2.4.15 Label variables

As with the values of the categories, sometimes it is desirable to have a longer, more descriptive variable name (see also formatting variables, 2.4.14). In general, we do not recommend using this feature, as it tends to complicate communication between data analysts and other readers of output (a possible exception is in graphical output).

```
data ds;
  ...
  label x="This is the label for the variable 'x'";
run;
```

The label is displayed instead of the variable name in all procedure output (except `proc print`, unless the `label` option is used) and can also be seen in `proc contents` (Section 2.3.1).

Some procedures also allow `label` statements with identical syntax, in which case the label is used only for that procedure.

2.4.16 Account for missing values

Example: See 2.13.3

Missing values are ubiquitous in most real-world investigations. The default missing value code for numeric data is '`.`', which has a numeric value of negative infinity. There are 27 other predefined missing value codes (`._`, `.a` ... `.z`), (listed in increasing numeric value) which can be used, for example, to record different reasons for missingness. The missing value code for character data, for assignment, is `" "` (quote blank quote), displayed as a blank.

Listwise deletion is usually the default behavior for most multivariate procedures. That is, observations with missing values for any variables named in the procedure are omitted from all calculations. Data step functions are different: functions defined with mathematical operators (`+ - / * **`) will result in a missing value if any operand has a missing value, but named functions, such as `sum(x1, x2)` will result in the function as applied to the non-missing values.

```
data ds;
  missing = (x1 eq .);
  x2 = (1 + 2 + .)/3;
  x3 = mean(1, 2, .);

  if x4 = 999 then x4 = .;

  x5 = n(1, 2, 49, 123, .);
  x6 = nmiss(x2, x3);

  if x1 ne .;
  if x1 ne . then output;
run;
```

The variable `missing` has a value of 1 if x_1 is missing, 0 otherwise. x_2 has a missing value, while $x_3 = 1.5$. Values of x_4 that were previously coded 999 are now marked as missing. The `n` function returns the number of non-missing values; x_5 has a value of 3; the `nmiss` function returns the number of missing values and here has a value of 1. The last two statements have identical meanings. They will remove all observations for which x_1 contains missing values.

2.4.17 Observation number

Example: See 2.13.2

```
data ...;
  obsnum = _n_;
run;
```

The variable `_n_` is created automatically by SAS and counts the number of lines of data that have been input into the `data` step. It is a temporary variable that it is not stored in the dataset unless a new variable is created (as demonstrated in the above code).

2.4.18 Unique values

Example: See 2.13.2

```
proc sort data=ds out=newds nodupkey;
  by x1 ... xk;
run;
```

The dataset `newds` contains all the variables in the dataset `ds`, but only one row for each unique value across $x_1 x_2 \ldots x_k$.

2.4.19 Lagged variable

A lagged variable has the value of that variable in a previous row (typically the immediately previous one) within that dataset. The value of lag for the first observation will be missing (see 2.4.16).

```
data ...;
  xlag1 = lag(x);
run;
```

or

```
data ...;
  xlagk = lagk(x);
run;
```

In the latter case, the variable `xlagk` contains the value of x from the kth preceding observation. The value of k can be any integer less than 101: the first `k` observations will have a missing value.

If executed conditionally, only observations with computed values are included. In other words, the statement `if (condition) then xlag1 = lag(x)` results in the variable `xlag1` containing the value of `x` from the most recently processed observation *for which* `condition` *was true*. This is a common cause of confusion.

2.4.20 SQL

Structured Query Language (SQL) is a language for querying and modifying databases. SAS supports access to SQL through `proc sql`.

2.4.21 Perl interface

Perl is a high-level general purpose programming language [44]. SAS 9.2 supports Perl regular expressions in the data step via the `prxparse`, `prxmatch`, `prxchange`, `prxparen`, and `prxposn` functions. Details on their use can be found in the online help: Contents; SAS Products; Base SAS; SAS 9.2 Language Reference:Dictionary; Functions and CALL Routines under the names listed above.

2.5 Merging, combining, and subsetting datasets

A common task in data analysis involves the combination, collation, and subsetting of datasets. In this section, we review these techniques for a variety of situations.

2.5.1 Subsetting observations

Example: See 2.13.4

```
data ...;
  if x eq 1;
run;
```

or

```
data ...;
  where x eq 1;
run;
```

or

```
data ...;
  set ds (where= (x eq 1));
run;
```

These examples create a new dataset consisting of observations where $x = 1$. The `if` statement has an implied "then output." The `where` syntax also works within procedures to limit the included observations to those that meet the condition, without creating a new dataset; see 5.6.9.

2.5.2 Random sample of a dataset

See also random number seed (2.10.9)

It is sometimes useful to sample a subset (here quantified as *nsamp*) of observations without replacement from a larger dataset.

```
data ds2;
  set ds;
  order = uniform(0);
run;

proc sort data=ds2;
  by order;
run;

data ds3;
  set ds2;
  if _n_ le nsamp;
run;
```

Note that after `proc sort` has run, `ds2` is a permuted version of `ds`, and that `ds3` is a subset of the permuted dataset.

It is also possible to generate a random sample in a single data step by generating a uniform random variate for each observation in the original data but using an `if` statement to retain only those which meet a criteria which changes with the number retained.

2.5.3 Convert from wide to long (tall) format

Example: See 5.6.9

Sometimes data are available in a different shape than that required for analysis. One example of this is commonly found in repeated longitudinal measures studies. In this setting it is convenient to store the data in a wide or multivariate format with one line per subject, containing typically subject invariant factors (e.g., gender), as well as a column for each repeated outcome. An example would be:

```
id female inc80 inc81 inc82
1 0 5000 5500 6000
2 1 2000 2200 3300
3 0 3000 2000 1000
```

where the income in 1980, 1981, and 1982 is included in one row for each id.

In contrast, SAS tools for repeated measures analyses (5.2.2) typically require a row for each repeated outcome, such as

```
id year female inc
1 80 0 5000
1 81 0 5500
1 82 0 6000
2 80 1 2000
2 81 1 2200
2 82 1 3300
3 80 0 3000
3 81 0 2000
3 82 0 1000
```

In this section and in Section 2.5.4 below, we show how to convert between these two forms of this example data.

```
data long;
  set wide;
  array incarray [3] inc80 - inc82;
  do year = 80 to 82;
    inc = incarray[year - 79];
    output;
  end;
  drop inc80 - inc82;
run;
```

or

```
data long;
  set wide;
  year=80; inc=inc80; output;
  year=81; inc=inc81; output;
  year=82; inc=inc82; output;
  drop inc80 - inc82;
run;
```

or

```
proc transpose data=wide out=long_i;
  var inc80 - inc82;
  by id female;
run;

data long;
  set long_i;
  year=substr(_name_, 4, 2)*1.0;
  drop _name_;
  rename col1=inc;
run;
```

The year=substr() statement in the last data step is required if the value of year must be numeric. The remainder of that step makes the desired variable name appear, and removes extraneous information.

2.5.4 Convert from long (tall) to wide format

See also Section 2.5.3 (reshape from wide to tall)

```
proc transpose data=long out=wide (drop=_name_) prefix=inc;
  var inc;
  id year;
  by id female;
run;
```

The (drop=_name_) option prevents the creation of an unneeded variable in the wide dataset.

2.5.5 Concatenate datasets

```
data newds;
  set ds1 ds2;
run;
```

The datasets ds1 and ds2 are assumed to previously exist. The newly created dataset newds has as many rows as the sum of rows in ds1 and ds2, and as many columns as unique variable names across the two input datasets.

2.5.6 Sort datasets

Example: See 2.13.4

```
proc sort data=ds;
  by x1 ... xk;
run;
```

The keyword descending can be inserted before any variable to sort that variable from high to low (see also 1.6.2).

2.5.7 Merge datasets

Example: See 5.6.11

Merging datasets is commonly required when data on single units are stored in multiple tables or datasets. We consider a simple example where variables `id`, `year`, `female` and `inc` are available in one dataset, and variables `id` and `maxval` in a second. For this simple example, with the first dataset given as:

```
id year female inc
1 80 0 5000
1 81 0 5500
1 82 0 6000
2 80 1 2000
2 81 1 2200
2 82 1 3300
3 80 0 3000
3 81 0 2000
3 82 0 1000
```

and the second given below.

```
id maxval
2 2400
1 1800
4 1900
```

The desired merged dataset would look like:

```
id year female inc maxval
1  81   0      5500 1800
1  80   0      5000 1800
1  82   0      6000 1800
2  82   1      3300 2400
2  80   1      2000 2400
2  81   1      2200 2400
3  82   0      1000 .
3  80   0      3000 .
3  81   0      2000 .
4  .    .      .    1900
```

```
proc sort data=ds1; by x1 ... xk;
run;

proc sort data=ds2; by x1 ... xk;
run;

data newds;
  merge ds1 ds2;
  by x1 ... xk;
run;
```

For example, the result desired in the note above can be created as follows, assuming the two datasets are named `ds1` and `ds2`:

```
proc sort data=ds1; by id; run;

proc sort data=ds2; by id; run;

data newds;
  merge ds1 ds2;
  by id;
run;
```

The `by` statement in the `data` step describes the matching criteria, in that every observation with a unique set of X_1 through X_k in `ds1` will be matched to every observation with the same set of X_1 through X_k in `ds2`. The output dataset will have as many columns as there are uniquely named variables in the input datasets, and as many rows as unique values across X_1 through X_k. The `by` statement can be omitted, which results in the nth row of each dataset contributing to the nth row of the output dataset, though this is rarely desirable. If matched rows have discrepant values for variables with the same name in multiple datasets, the value in the later-named dataset is used in the output dataset.

It is sometimes useful to know whether an observation in the merged dataset is coming from one or more of the constituent datasets. This can be done using the `in=varname` dataset option. Using the syntax `(in=varname)` immediately following the dataset name in the `merge` statement causes the `varname` variable to be created with a value 1 if the observation is in the dataset and 0 otherwise. For example, consider the following code, as applied to the above example datasets:

```
data newds2;
  merge ds1 (in=inds1) ds2 (in=inds2);
  by id;
  if inds1 and inds2;
run;
```

The output dataset `newds2` will contain only observations which appear in both input datasets, meaning in this case that all observations on IDs 3 and 4 will be omitted. The `varname` variables are not included in the output data-set.

2.5.8 Drop variables in a dataset

Example: See 2.13.1

It is often desirable to prune extraneous variables from a dataset to simplify analyses.

```
data ds;
  ...
  keep x1 xk;
  ...
run;
```

or

```
data ds;
  set old_ds (keep=x1 xk);
  ...
run;
```

The complementary syntax `drop` can also be used, both as a statement in the data step and as a `data` statement option.

2.6 Date and time variables

Variables in the date formats are integers counting the number of days since January 1, 1960. Time variables are integers counting the number of seconds since midnight, December 31, 1959 or since the most recent midnight.

2.6.1 Create date variable

```
data ...;
  dayvar = input("04/29/2010", mmddyy10.);
  todays_date = today();
run;
```

The variable `dayvar` contains the integer number of days between January 1, 1960 and April 29, 2010. The value of `todays_date` is the integer number of days between January 1, 1960 and the day the current instance of SAS was opened.

2.6.2 Extract weekday

```
data ...;
  wkday = weekday(datevar);
run;
```

The `weekday` function returns an integer representing the weekday, 1=Sunday, ..., 7=Saturday.

2.6.3 Extract month

```
data ...;
  monthval = month(datevar);
run;
```

The `month` function returns an integer representing the month, 1=January, ..., 12=December.

2.6.4 Extract year

```
data ...;
  yearval = year(datevar);
run;
```

The variable `yearval` is years counted in the Common Era (CE, also called AD).

2.6.5 Extract quarter

```
data ...;
  qtrval = qrt(datevar);
run;
```

The return values for `qtrval` are 1, 2, 3, or 4.

2.6.6 Extract time from a date-time value

```
data ...;
  timevar = timepart(datetimevar);
run;
```

The new variable `timevar` contains the number of seconds after midnight, regardless of what date is included in `datetimevar`.

2.6.7 Create time variable

See also 2.7.1 (timing commands)

```
data ...;
  timevar_1960 = datetime();
  timevar_midnight = time();
run;
```

The variable `timevar_1960` contains the number of seconds since midnight, December 31, 1959. The variable `timevar_midnight` contains the number of seconds since the most recent midnight.

2.7 Interactions with the operating system

2.7.1 Timing commands

```
options stimer;
options fullstimer;
```

These options request that a subset (`stimer`) or all available (`fullstimer`) statistics are reported in the SAS log.

2.7.2 Execute command in operating system

```
x;
```

or

```
x 'OS command';
```

or

```
data ...;
  call system("OS command");
run;
```

An example command statement would be `x 'dir'`. The statement consisting of just `x` will open a command window. Related statements are `x1`, `x2`, ..., `x9`, which allow up to 9 separate operating system tasks to be executed simultaneously.

The `x` command need not be in a data step, and cannot be executed conditionally. In other words, if it appears as a consequence in an `if` statement, it will be executed regardless of whether or not the test in the `if` statement is true or not. Use the `call system` statement as shown to execute conditionally.

This syntax to open a command window may not be available in all operating systems.

2.7.3 Find working directory

```
x;
```

This will open a command window; the current directory in this window is the working directory. The working directory can also be found using the method shown in Section 2.7.5 using the `cd` command in Windows or the `pwd` command in Linux.

The current directory is displayed by default in the status line at the bottom of the SAS window.

2.7.4 Change working directory

```
x 'cd dir_location';
```

This can also be done interactively by double-clicking the display of the current directory in the status line at the bottom of the SAS window (note that this applies for Windows installations, for other operating systems, see the online help: Contents; Using SAS software in Your Operating Environment; SAS 9.2 companion for <your OS>; Running SAS under <your OS>).

2.7.5 List and access files

```
filename filehandle pipe 'dir /b'; /* Windows */
filename filehandle pipe 'ls'; /* Unix or Mac OS X */

data ds;
  infile filehandle truncover;
  input x $20.;
run;
```

The `pipe` is a special file type which passes the characters inside the single quote to the operating system when read using the `infile` statement, then reads the result. The above code lists the contents of the current directory. The dataset `ds` contains a single character variable `x` with the file names. The file handle can be no longer than eight characters.

2.8 Mathematical functions

2.8.1 Basic functions

```
data ...;
  minx = min(x1, ..., xk);
  maxx = max(x1, ..., xk);
  meanx = mean(x1, ..., xk);
  stddevx = std(x1, ..., xk);
  sumx = sum(x1, ..., xk)
  absolutevaluex = abs(x);
  etothex = exp(x);
  xtothey = x**y;
  squareroottx = sqrt(x);
  naturallogx = log(x);
  logbase10x = log10(x);
  logbase2x = log2(x);
  remainder = mod(x1, x2);
run;
```

The first five functions operate on a row-by-row basis. The last function returns the remainder when dividing x_1 by x_2.

2.8.2 Trigonometric functions

```
data ...;
  sinx = sin(x);
  sinpi = sin(constant('PI'));
  cosx = cos(x);
  tanx = tan(x);
  arccosx = arcos(x);
  arcsinx = arsin(x);
  arctanx = atan(x);
  arctanxy = atan2(x, y);
run;
```

2.8.3 Special functions

```
data ...;
  betaxy = beta(x, y);
  gammax = gamma(x);
  factorialn = fact(n);
  nchooser = comb(n, r);
  npermuter = perm(n, r);
run;
```

2.8.4 Integer functions

See also 2.2.6 (rounding and number of digits to display)

```
data ...;
  nextintx = ceil(x);
  justintx = floor(x);
  roundx = round(x1, x2);
  roundint = round(x, 1);
  movetozero = int(x);
run;
```

The value of `roundx` is X_1, rounded to the nearest X_2. The value of `movetozero` is the same as `justint` if $x > 0$ or `nextint` if $x < 0$.

2.8.5 Comparisons of floating point variables

Because certain floating point values of variables do not have exact decimal equivalents, there may be some error in how they are represented on a computer. For example, if the true value of a particular variable is 1/7, the approximate decimal is 0.1428571428571428. For some operations (for example, tests of equality), this approximation can be problematic.

```
data ds;
  x1 = ((1/7) eq .142857142857);
  x2 = (fuzz((1/7) - .142857142857) eq 0);
run;
```

In the above example, $x_1 = 0$, $x_2 = 1$. If the argument to `fuzz` is less than 10^{-12}, then the result is the nearest integer.

2.8.6 Optimization problems

SAS can be used to solve optimization (maximization) problems. As an extremely simple example, consider maximizing the area of a rectangle with perimeter equal to 20. Each of the sides can be represented by x and 10-x, with area of the rectangle equal to $x * (10 - x)$.

```
proc iml;
  start f_area(x);
  f = x*(10-x);
  return (f);
  finish f_area;
  con = {0, 10};
  x = {2} ;
  optn = {1, 2};
  call nlpcg(rc, xres, "f_area", x, optn, con);
quit;
```

The above uses conjugate gradient optimization. Several additional optimization routines are provided in proc iml (see the online help: Contents; SAS Products; SAS/IML User's Guide; Nonlinear Optimization Examples).

2.9 Matrix operations

Matrix operations are often needed in statistical analysis. The SAS/IML product (separate from SAS/STAT), includes proc iml, which is used to treat data as a matrix.

Here, we briefly outline the process needed to read a SAS dataset into SAS/IML as a matrix, perform some function, then make the result available as a SAS native dataset. Throughout this section, we use capital letters to emphasize that a matrix is described, though proc iml is not case-sensitive.

```
proc iml;
  use ds;
  read all var{x1 ... xk} into Matrix_x;
  ... /* perform a function of some sort */
  print Matrix_x; /* print the matrix to the output window */
  create newds from Matrix_x;
  append from Matrix_x;
quit;
```

Calls to proc iml end with a quit statement, rather than a run statement.

2.9.1 Create matrix directly

In this entry, we demonstrate creating a 2×2 matrix consisting of the first four nonzero integers:

$$A = \begin{pmatrix} 1 & 2 \\ 3 & 4 \end{pmatrix}.$$

```
proc iml;
A = {1 2, 3 4};
quit;
```

2.9.2 Create matrix by combining matrices

```
proc iml;
  A = {1 2, 3 4};
  B = {1 2, 3 4};
  AaboveB = A//B;
  BrightofA = A||B;
quit;
```

The new matrix AaboveB is a 4×2 matrix; BrightofA is a 2×4 matrix.

2.9.3 Transpose matrix

```
proc iml;
  A = {1 2, 3 4};
  transA = A`;
  transA_2 = t(A);
quit;
```

Both transA and transA_2 contain the transpose of A.

2.9.4 Matrix multiplication

```
proc iml;
  A = {1 2, 3 4};
  B = {1 2, 3 4};
  ABm = A*B;
quit;
```

2.9.5 Elementwise multiplication

```
proc iml;
  A = {1 2, 3 4};
  B = {1 2, 3 4};
  ABe = A#B;
quit;
```

2.9.6 Invert matrix

```
proc iml;
  A = {1 2, 3 4};
  Ainv = inv(A);
quit;
```

2.9.7 Create submatrix

```
proc iml;
  A = {1 2 3 4, 5 6 7 8, 9 10 11 12};
  Asub = a[2:3, 3:4];
quit;
```

2.9.8 Create a diagonal matrix

```
proc iml;
  A = {1 2, 3 4};
  diagMat = diag(A);
quit;
```

For matrix A, this results in a matrix with the same diagonals, but with all off-diagonals set to 0. For vector argument, the function generates a matrix with the vector values as the diagonals and all off-diagonals 0.

2.9.9 Create vector of diagonal elements

```
proc iml;
  A = {1 2, 3 4};
  diagVals = vecdiag(A);
quit;
```

The vector `diagVals` contains the diagonal elements of matrix `A`.

2.9.10 Create vector from a matrix

```
proc iml;
  A = {1 2, 3 4};
  newvec = shape(A, 1);
quit;
```

This makes a row vector from all the values in the matrix.

2.9.11 Calculate determinant

```
proc iml;
  A = {1 2, 3 4};
  detval = det(A);
quit;
```

2.9.12 Find eigenvalues and eigenvectors

```
proc iml;
  A = {1 2, 3 4};
  Aeval = eigval(A);
  Aevec = eigvec(A);
quit;
```

2.9.13 Calculate singular value decomposition

The singular value decomposition of a matrix A is given by $A = U * \text{diag}(Q) * V^T$ where $U^TU = V^TV = VV^T = I$ and Q contains the singular values of A.

```
proc iml;
  A = {1 2, 3 4};
  call svd(U, Q, V, A);
quit;
```

2.10 Probability distributions and random number generation

SAS can calculate quantiles and cumulative distribution values as well as generate random numbers for a large number of distributions. Random variables are commonly needed for simulation and analysis. Comprehensive random number generation is provided by the rand function.

A seed can be specified for the random number generator. This is important to allow replication of results (e.g., while testing and debugging). Information about random number seeds can be found in Section 2.10.9.

Table 2.1 summarizes support for quantiles, cumulative distribution functions, and random number generation.

2.10.1 Probability density function

Similar syntax is used for a variety of distributions. Here we use the normal distribution as an example; others are shown in Table 2.1.

```
data ...;
  y = cdf('NORMAL', 1.96, 0, 1);
run;
```

2.10.2 Quantiles of a probability density function

Similar syntax is used for a variety of distributions. Here we use the normal distribution as an example; others are shown in Table 2.1.

```
data ...;
  y = quantile('NORMAL', .975, 0, 1);
run;
```

2.10.3 Uniform random variables

```
data ...;
  x1 = uniform(seed);
  x2 = rand('UNIFORM');
run;
```

The variables x_1 and x_2 are uniform on the interval (0,1). The ranuni() function is a synonym for uniform().

Table 2.1: Quantiles, Probabilities, and Pseudorandom Number Generation: Distributions Available in SAS

Distribution	SAS DISTNAME
Beta	BETA
binomial	BINOMIAL
Cauchy	CAUCHY
chi-square	CHISQUARE
exponential	EXPONENTIAL
F	F
gamma	GAMMA
geometric	GEOMETRIC
hypergeometric	HYPERGEOMETRIC
inverse normal	IGAUSS*
Laplace	LAPLACE
logistic	LOGISTIC
lognormal	LOGNORMAL
negative binomial	NEGBINOMIAL
normal	NORMAL
Poisson	POISSON
Student's t	T
Uniform	UNIFORM
Weibull	WEIBULL

Note: Random variates can be generated from the `rand` function: `rand('DISTNAME', parm1, ..., parmn)`, the areas to the left of a value via the `cdf` function: `cdf('DISTNAME', quantile, parm1, ..., parmn)`, and the quantile associated with a probability (the inverse CDF) via the `quantile` function: `quantile('DISTNAME', probability, parm1, ..., parmn)`, where the number of `parms` varies by distribution. Details are available through the online help: Contents; SAS Products; Base SAS; SAS 9.2 Language Reference: Dictionary; Dictionary of Language Elements; Functions and CALL Routines; RAND Function. Note that in this instance SAS is case-sensitive.
*The inverse normal is not available in the `rand` function; inverse normal variates can be generated by taking the inverse of normal random variates.

2.10.4 Multinomial random variables

```
data ...;
  x1 = rantbl(seed, p1, p2, ..., pk);
  x2 = rand('TABLE', p1, p2, ..., pk);
run;
```

The variables x_1 and x_2 take the value i with probability p_i and value $k+1$ with value $1 - \sum_{i=1}^{k} p_i$.

2.10.5 Normal random variables

Example: See 2.13.5

```
data ...;
  x1 = normal(seed);
  x2 = rand('NORMAL', mu, sigma);
run;
```

The variable X_1 is a standard normal ($\mu = 0$ and $\sigma = 1$), while X_2 is normal with specified mean and standard deviation. The function `rannor()` is a synonym for `normal()`.

2.10.6 Multivariate normal random variables

For the following, we first create a 3×3 covariance matrix. Then we generate 1,000 realizations of a multivariate normal vector with the appropriate correlation or covariance.

```
data Sigma (type=cov);
  infile cards;
  input _type_ $ _Name_ $ x1 x2 x3;
  cards;
cov x1 3 1 2
cov x2 1 4 0
cov x3 2 0 5
;
run;

proc simnormal data=sigma out=outtest2 numreal=1000;
  var x1 x2 x3;
run;
```

The `type=cov` option to the `data` step defines `Sigma` as a special type of SAS dataset which contains a covariance matrix in the format shown. A similar `type=corr` dataset can be used to generate using a correlation matrix instead of a covariance matrix.

2.10.7 Exponential random variables

```
data ...;
  x1 = ranexp(seed);
  x2 = rand('EXPONENTIAL');
run;
```

The expected value of both X_1 and X_2 is 1: for exponentials with expected value k, multiply the generated value by k.

2.10.8 Other random variables

Example: See 2.13.5

The list of probability distributions supported within SAS can be found in Table 2.1. In addition to these distributions, the inverse probability integral transform can be used to generate arbitrary random variables with invertible cumulative density function F (exploiting the fact that $F^{-1} \sim U(0,1)$). As an example, consider the generation of random variates from an exponential distribution with rate parameter λ, where $F(X) = 1 - \exp(-\lambda X) = U$. Solving for X yields $X = -\log(1-U)/\lambda$. If we generate a uniform(0,1) variable, we can use this relationship to generate an exponential with the desired rate parameter.

```
data ds;
  lambda = 2;
  uvar = uniform(42);
  expvar = -1 * log(1-uvar)/lambda;
run;
```

2.10.9 Setting the random number seed

Comprehensive random number generation is provided by the `rand` function. For variables created this way, an initial seed is selected automatically based on the system clock. Sequential calls use a seed derived from this initial seed. To generate a replicable series of random variables, use the `call streaminit` function before the first call to `rand`.

```
call streaminit(42);
```

A set of separate functions for random number generation includes `normal`, `ranbin`, `rancau`, `ranexp`, `rangam`, `rannor`, `ranpoi`, `rantbl`, `rantri`, `ranuni`, and `uniform`. For these functions, calling with an argument of (0) is equivalent to calling the `rand` function without first running `call streaminit`; an initial seed is generated from the system clock. Calling the same functions with an integer greater than 0 as argument is equivalent to running `call streaminit` before an initial use of `rand`. In other words, this will result in a series of variates based on the first specified integer. Note that `call streaminit` or specifying an integer to one of the specific functions need only be performed once per `data` step; all seeds within that `data` step will be based on that seed.

2.11 Control flow, programming, and data generation

Here we show some key aspects of programming.

2.11.1 Looping

Example: See 7.2.2

```
data;
  do i = i1 to i2;
    x = normal(0);
    output;
  end;
run;
```

The above code generates a new dataset with $i_2 - i_1 + 1$ standard normal variates, with seed based on the system clock (2.10.5). The generic syntax for looping includes three parts: (1) a `do varname = val1 to val2` statement; (2) the statements to be executed within the loop; and (3) an `end` statement. As with all programming languages, users should be careful about modifying the index during processing. Other options include `do while` and `do until`. To step values of `i` by values other than 1, use statements such as `do i = i1 to i2 by byval`. To step across specified values, use statements like `do k1, ... , kn`.

2.11.2 Conditional execution

Example: See 2.13.3

```
data ds;
  if expression1 then expression2 else expression3;
run;
```

or

```
if expression1 then expression2;
else if expression3 then expression4;
...
else expressionk;
```

or

```
if expression1 then do;
  ...;
end;
else if expression2 then expression3;
...
```

There is no limit on the number of conditions tested in the `else` statements, which always refer back to the most recent `if` statement. Once a condition in this sequence is met, the remaining conditions are not tested. Listing conditions in decreasing order of occurrence will therefore result in more efficient code.

The `then` code is executed if the `expression` following the `if` has a non-missing, nonzero value. So, for example, the statement `if 1 then y = x**2` is valid syntax, equivalent to the statement `y=x**2`. Good programming style is to make each tested expression be a logical test, such as `x eq 1` returning 1 if the expression is true and 0 otherwise. SAS includes mnemonics `lt`, `le`, `eq`, `ge`, `gt`, and `ne` for $<$, $\leq$, $=$, $\geq$, $>$, and $\neq$, respectively. The mnemonic syntax cannot be used for assignment, and it is recommended style to reserve `=` for assignment and use only the mnemonics for testing.

The `do-end` block is all executed conditionally. Any group of `data` step statements can be included in a `do-end` block.

2.11.3 Sequence of values or patterns

Example: See 2.13.5

It is often useful to generate a variable consisting of a sequence of values (e.g., the integers from 1 to 100) or a pattern of values (1 1 1 2 2 2 3 3 3). This might be needed to generate a variable consisting of a set of repeated values for use in a simulation or graphical display.

```
data ds;
  do x = 1 to nvals;
    ...
  end;
run;
```

As an example, we demonstrate generating data from a linear regression model of the form:

$$E[Y|X_1, X_2] = \beta_0 + \beta_1 X_1 + \beta_2 X_2, \; Var(Y|X) = 3, \; Corr(X_1, X_2) = 0.$$

The following code implements the model described above for $n = 200$. The value 42 below is an arbitrary seed (2.10.9) [1] used for random number generation. The datasets `ds1` and `ds2` will be identical. However such values are generated, it would be wise to use `proc freq` (3.3.1) to check whether the intended results were achieved.

```
data ds1;
  beta0 = -1; beta1 = 1.5; beta2 = .5; mse = 3;
  /* note multiple statements on previous line */
  do x1 = 1 to 2;
    do x2 = 1 to 2;
      do obs = 1 to 50;
        y = beta0 + beta1*x1 + beta2*x2 + normal(42)*mse;
        output;
      end;
    end;
  end;
run;
```

or

```
data ds2;
  beta0 = -1; beta1 = 1.5; beta2 = .5; mse = 3;
  do i = 1 to 200;
    x1 = (i gt 100) + 1;
    x2 = (((i gt 50) and (i le 100)) or (i gt 150)) + 1;
    y = beta0 + beta1*x1 + beta2*x2 + normal(42)*mse;
    output;
  end;
run;
```

2.11.4 Referring to a range of variables

Example: See 2.13.3

For functions such as `mean()` it is often desirable to list variables to be averaged without listing them all by name. SAS provides two ways of doing this. First, variables stored adjacently can be referred to as a range `vara -- varb` (with two hyphens). Inserting `numeric` or `character` between the hyphens includes only variables of that type, so that `vara -numeric- varb` includes all numeric variables stored between `vara` and `varb`, inclusive. Variables with sequential numerical suffices can be referred to as a range `varname1 - varnamek` (with a single hyphen) regardless of the storage location. The key thing to bear in mind is that the part of the name before the number must be identical for all variables. Finally the colon (:) works as a wildcard. Thus "string:" includes all variables that start with "string". All versions of this shorthand syntax also work in procedures.

```
data ...;
  meanadjacentx = mean(of x1 -- xk);
  mean_numeric_x = mean(of x1 -numeric- xk);
  meannamedx = mean(of x1 - xk);
  mean_all_x = mean(of x:);
run;
```

The first code will return the mean of all the variables stored between x_1 and x_k. The second will include numeric variables stored between those variables. The third will return the mean of $x_1 \ldots x_k$, if they all exist, and will create any variables in the range if they do not exist. The final line produces the mean of all variables whose names begin with "x".

2.11.5 Perform an action repeatedly over a set of variables

Example: See 2.13.3 and 5.6.9

It is often necessary to perform a given function for a series of variables. Here the square of each of a list of variables is calculated as an example. This can be accomplished using arrays.

```
data ...;
  array arrayname1 [arraylen] x1 x2 ... xk;
  array arrayname2 [arraylen] z1 ... zk;
  do i = 1 to arraylen;
    arrayname2[i] = arrayname1[i]**2;
  end;
run;
```

In the above example, $z_i = x_i^2, i = 1 \ldots k$, for every observation in the dataset. The variable `arraylen` is an integer. It can be replaced by '*', which implies that the dimension of the array is to be calculated automatically by SAS from the number of elements. Elements (variables in the array) are listed after the brackets. Arrays can also be multidimensional, when multiple dimensions are specified (separated by commas) within the brackets. This can be useful, for example, when variables contain a matrix for each observation in the dataset.

Variables can be created by definition in the array statement, meaning that in the above code, the variable `x2` need not exist prior to the first `array` statement. The function `dim(arrayname1)` returns the number of elements in the array, and can be used in place of the variable `arraylen` to loop over arrays declared with the '*' syntax.

2.12 Further resources

Comprehensive introductions to data management in SAS can be found in Delwiche and Slaughter [9] and Cody and Smith [4]. Paul Murrell's forthcoming *Introduction to Data Technologies* text [32] provides a comprehensive introduction to XML, SQL, and other related technologies and can be found at `http://www.stat.auckland.ac.nz/~paul/ItDT`.

2.13 HELP examples

To help illustrate the tools presented in this chapter, we apply many of the entries to the HELP data. SAS code can be downloaded from `http://www.math.smith.edu/sas/examples`.

2.13.1 Data input and output

We begin by reading the dataset (2.1.4), keeping only the variables that are needed (2.5.8).

```
proc import
  datafile='c:/book/help.csv' out=dsprelim dbms=dlm;
  delimiter=',';
  getnames=yes;
run;

data ds;
  set dsprelim;
  keep id cesd f1a -- f1t i1 i2 female treat;
run;
```

Here `proc import` reads the `csv` file from the named location on the hard disk, while the `data` step selects the variables. Note the use of the '--' syntax (2.11.4) to keep all of the variables stored from `f1a` through `f1t`. We can then show a summary of the dataset. We use the `ODS` system (1.7) to reduce the length of the output. We begin by examining the `attributes` output.

```
options ls=64;   /* narrows width to stay in grey box */
ods select attributes;
proc contents data=ds;
run;
ods select all;

The CONTENTS Procedure
Data Set Name          WORK.DS            Observations          453
Member Type            DATA               Variables             26
Engine                 V9                 Indexes               0
Created                Wed, Feb 10,       Observation Length    208
                       2010 09:46:35 PM
Last Modified          Wed, Feb 10,       Deleted Observations  0
                       2010 09:46:35 PM
Protection                                Compressed            NO
Data Set Type                             Sorted                NO
Label
Data Representation    WINDOWS_32
Encoding               wlatin1  Western
                       (Windows)
```

The default output (without selecting pieces with ODS commands) prints another dataset with a line for each variable. This shows its name and additional information; the short option below limits the output to just the names of the variable.

```
options ls=64;   /* narrows width to stay in grey box */
ods select variablesshort;
proc contents data=ds short;
run;
ods select all;

The CONTENTS Procedure
         Alphabetic List of Variables for WORK.DS
cesd f1a f1b f1c f1d f1e f1f f1g f1h f1i f1j f1k f1l f1m f1n f1o
f1p f1q f1r f1s f1t female i1 i2 id treat
```

Displaying the first few rows of data can give a more concrete sense of what is in the dataset:

```
proc print data=ds (obs=5) width=minimum;
run;

                                                        f
                                                   t    e
                                                   r  c m
 O   f f f f f f f f f f f f f f f f f f f f       e  e a
 b i 1 1 1 1 1 1 1 1 1 1 1 1 1 1 1 1 1 1 1 1  i  i a  s l
 s d a b c d e f g h i j k l m n o p q r s t  1  2 t  d e
 1 1 3 2 3 0 2 3 3 0 2 3 3 0 1 2 2 2 2 3 3 2 13 26 1 49 0
 2 2 3 2 0 3 3 2 0 0 3 0 3 0 0 3 0 0 0 2 0 0 56 62 1 30 0
 3 3 3 2 3 0 2 2 1 3 2 3 1 0 1 3 2 0 0 3 2 0  0  0 0 39 0
 4 4 0 0 1 3 2 2 1 3 0 0 1 2 2 2 0 . 2 0 0 1  5  5 0 15 1
 5 5 3 0 3 3 3 3 1 3 3 2 3 2 2 3 0 3 3 3 3 3 10 13 0 39 0
```

Saving the dataset in native format (2.2.1) will simplify future access. We also add a comment (2.3.4) to help later users understand what is in the dataset.

```
libname book 'c:/temp';
data book.ds (label = "HELP baseline dataset");
  set ds;
run;
```

Saving it in a foreign format (2.1.5), say Microsoft Excel, will allow access to other tools for analysis and display:

```
proc export data=ds replace
  outfile="c:/temp/ds.xls"
  dbms=excel;
run;
```

The `replace` option above is available in many settings where SAS can save a file in the operating system. It will replace any existing file with the named file. If `replace` is omitted and the file (`c:/temp/ds.xls` in this case) already exists, an error message will be generated.

2.13.2 Data display

We begin by consideration of the CESD (Center for Epidemiologic Studies-Depression) measure of depressive symptoms for this sample at baseline.

```
proc print data=ds (obs=10);
  var cesd;
run;

Obs         cesd
  1           49
  2           30
  3           39
  4           15
  5           39
  6            6
  7           52
  8           32
  9           50
 10           46
```

It may be useful to know how many high values there are, and to which observations they belong:

```
proc print data=ds;
  where cesd gt 55;
  var cesd;
run;

Obs         cesd
 64           57
116           58
171           57
194           60
231           58
266           56
295           58
305           56
387           57
415           56
```

Similarly, it may be useful to examine the observations with the lowest values:

```
options ls=64;
proc sort data=ds out=dss1;
  by cesd;
run;

proc print data=dss1 (obs=4);
  var id cesd i1 treat;
run;

Obs          id          cesd          i1          treat
  1         233             1           3              0
  2         418             3          13              0
  3         139             3           1              0
  4          95             4           9              1
```

2.13.3 Derived variables and data manipulation

Suppose the dataset arrived with only the individual CESD questions, and not the sum. We would need to create the CESD score. We will do this using an array (2.11.5) to aid the recoding of the four questions which are asked "backwards," meaning that high values of the response are counted for fewer points.[1] To demonstrate other tools, we will also see if there is any missing data (2.4.16), and calculate the score using the mean of the observed values. This is equivalent to imputing the mean of the observed values for any missing values.

```
data cesd;
  set ds;
  /* list of backwards questions */
  array backwards [*] f1d f1h f1l f1p;
  /* for each, subtract the stored value from 3 */
  do i = 1 to dim(backwards);
     backwards[i] = 3 - backwards[i];
  end;
  /* this generates the sum of the non-missing questions */
  newcesd = sum(of f1a -- f1t);
  /* This counts the number of missing values, per person */
  nmisscesd = nmiss(of f1a -- f1t);
  /* this gives the sum, imputing the mean of non-missing */
  imputemeancesd = mean(of f1a -- f1t) * 20;
run;
```

We will check our CESD score against the one which came with the dataset.

[1] According to the coding instructions found at `http://patienteducation.stanford.edu/research/cesd.pdf`.

To evaluate the missing data approach, we print the subjects with any missing questions.

```
proc print data=cesd (obs=20);
  where nmisscesd gt 0;
  var cesd newcesd nmisscesd imputemeancesd;
run;

Obs          cesd     newcesd     nmisscesd     imputemeancesd
  4            15          15             1            15.7895
 17            19          19             1            20.0000
 87            44          44             1            46.3158
101            17          17             1            17.8947
154            29          29             1            30.5263
177            44          44             1            46.3158
229            39          39             1            41.0526
```

The output shows that the original dataset was created with unanswered questions counted as if they had been answered with a zero. This conforms to the instructions provided with the CESD, but might be questioned on theoretical grounds.

It is often necessary to create a new variable using logic (2.4.13). In the HELP study, many subjects reported extreme amounts of drinking. Here, an ordinal measure of alcohol consumption (abstinent, moderate, high-risk) is created using information about average consumption per day in the past 30 days prior to detox (`i1`, measured in standard drink units) and maximum number of drinks per day in the past 30 days prior to detox (`i2`). The number of drinks required for each category differ for men and women according to National Institute of Alcohol Abuse and Alcoholism (NIAAA) guidelines for physicians [33].

```
data ds2;
set ds;
  if i1 eq 0 then drinkstat="abstinent";
  if (i1 eq 1 and i2 le 3 and female eq 1) or
    (((i1 eq 1) or (i1 eq 2)) and i2 le 4 and female eq 0)
    then drinkstat="moderate";
  if (((i1 gt 1) or (i2 gt 3)) and female eq 1) or
    (((i1 gt 2) or (i2 gt 4)) and female eq 0)
    then drinkstat="highrisk";
  if nmiss(i1,i2,female) ne 0 then drinkstat="";
run;
```

It is always prudent to check the results of derived variables. As a demonstration, we display the observations in the 361st through 370th rows of the data.

```
proc print data=ds2 (firstobs=361 obs=370);
  var i1 i2 female drinkstat;
run;

Obs             i1              i2          female    drinkstat
361             37              37               0    highrisk
362             25              25               0    highrisk
363             38              38               0    highrisk
364             12              29               0    highrisk
365              6              24               0    highrisk
366              6               6               0    highrisk
367              0               0               0    abstinent
368              0               0               1    abstinent
369              8               8               0    highrisk
370             32              32               0    highrisk
```

It is also useful to focus such checks on a subset of observations. Here we show the drinking data for moderate female drinkers.

```
proc print data=ds2;
  where drinkstat eq "moderate" and female eq 1;
  var i1 i2 female drinkstat;
run;

Obs             i1              i2          female    drinkstat
116              1               1               1    moderate
137              1               3               1    moderate
225              1               2               1    moderate
230              1               1               1    moderate
264              1               1               1    moderate
266              1               1               1    moderate
394              1               1               1    moderate
```

Basic data description is an early step in analysis. Here we show some summary statistics related to drinking and gender.

```
proc freq data=ds2;
  tables drinkstat;
run;

The FREQ Procedure
                                          Cumulative    Cumulative
drinkstat    Frequency      Percent       Frequency       Percent
--------------------------------------------------------------
abstinent           68        15.01              68         15.01
highrisk           357        78.81             425         93.82
moderate            28         6.18             453        100.00
```

```
proc freq data=ds2;
  tables drinkstat*female;
run;

The FREQ Procedure
Table of drinkstat by female
drinkstat     female
Frequency |
Percent   |
Row Pct   |
Col Pct   |       0|       1|  Total
----------+--------+--------+
abstinent |     42 |     26 |     68
          |   9.27 |   5.74 |  15.01
          |  61.76 |  38.24 |
          |  12.14 |  24.30 |
----------+--------+--------+
highrisk  |    283 |     74 |    357
          |  62.47 |  16.34 |  78.81
          |  79.27 |  20.73 |
          |  81.79 |  69.16 |
----------+--------+--------+
moderate  |     21 |      7 |     28
          |   4.64 |   1.55 |   6.18
          |  75.00 |  25.00 |
          |   6.07 |   6.54 |
----------+--------+--------+
Total          346      107      453
             76.38    23.62   100.00
```

To display gender in a more direct fashion, we create a new character variable. Note that in the following quoted strings, SAS is case-sensitive.

```
data ds3;
set ds;
  if female eq 1 then gender="Female";
  else if female eq 0 then gender="male";
run;

proc freq data=ds3;
  tables female gender;
run;

The FREQ Procedure
                                        Cumulative    Cumulative
female    Frequency     Percent      Frequency       Percent
-------------------------------------------------------------
     0         346       76.38            346         76.38
     1         107       23.62            453        100.00

                                        Cumulative    Cumulative
gender    Frequency     Percent      Frequency       Percent
-------------------------------------------------------------
Female         107       23.62            107         23.62
male           346       76.38            453        100.00
```

2.13.4 Sorting and subsetting datasets

It is often useful to sort datasets (2.5.6) by the order of a particular variable (or variables). Here we sort by CESD and drinking.

```
proc sort data=ds;
  by cesd i1;
run;

proc print data=ds (obs=5);
  var id cesd i1;
run;

Obs            id          cesd            i1
  1           233             1             3
  2           139             3             1
  3           418             3            13
  4           251             4             4
  5            95             4             9
```

It is sometimes necessary to create a dataset that is a subset (2.5.1) of another dataset. For example, here we make a dataset which only includes female subjects. First, we create the subset and calculate a summary value in the resulting dataset.

```
data females;
set ds;
  where female eq 1;
run;

proc means data=females mean maxdec=1;
  var cesd;
run;

The MEANS Procedure
Analysis Variable : cesd

        Mean
------------
        36.9
------------
```

To test the subsetting, we display the mean for both genders, using the original dataset.

```
proc sort data=ds;
  by female;
run;

proc means data=ds mean maxdec=2;
  by female;
  var cesd;
run;

female=0
The MEANS Procedure
Analysis Variable : cesd

        Mean
------------
       31.60
------------

female=1
Analysis Variable : cesd

        Mean
------------
       36.89
------------
```

The ODS system (1.7) provides a way to save the two means in a dataset.

```
ods exclude all;
ods output summary=means;
proc means data=ds mean maxdec=2;
  by female;
  var cesd;
run;
ods select all;

proc print data=means;
run;

Obs            female          cesd_Mean
 1                0               31.60
 2                1               36.89
```

2.13.5 Probability distributions

To demonstrate more tools, we leave the HELP dataset and show examples of how data can be generated. We will generate values (2.10.5) from the normal and t distribution densities.

```
data dists;
  do x = -4 to 4 by .1;
    normal_01 = sqrt(2 * constant('PI'))**(-1) *
      exp(-1 * ((x*x)/2)) ;
    dfval = 1;
    t_1df = (gamma((dfval +1)/2) / (sqrt(dfval *
      constant('PI')) * gamma(dfval/2))) *
      (1 + (x*x)/dfval)**(-1 * ((dfval + 1)/2));
    output;
  end;
run;
```

Figure 2.4 displays a plot of these distributions.

```
legend1 label=none position=(top inside right) frame down=2
   value = ("N(0,1)" tick=2 "t with 1 df");
axis1 label=(angle=90 "f(x)") minor=none order=(0 to .4 by .1);
axis2 minor=none order=(-4 to 4 by 2);
symbol1 i=j v=none l=1 c=black w=5;
symbol2 i=j v=none l=21 c=black w=5;
proc gplot data= dists;
  plot (normal_01 t_1df) * x / overlay legend=legend1
    vaxis=axis1 haxis=axis2;
run; quit;
```

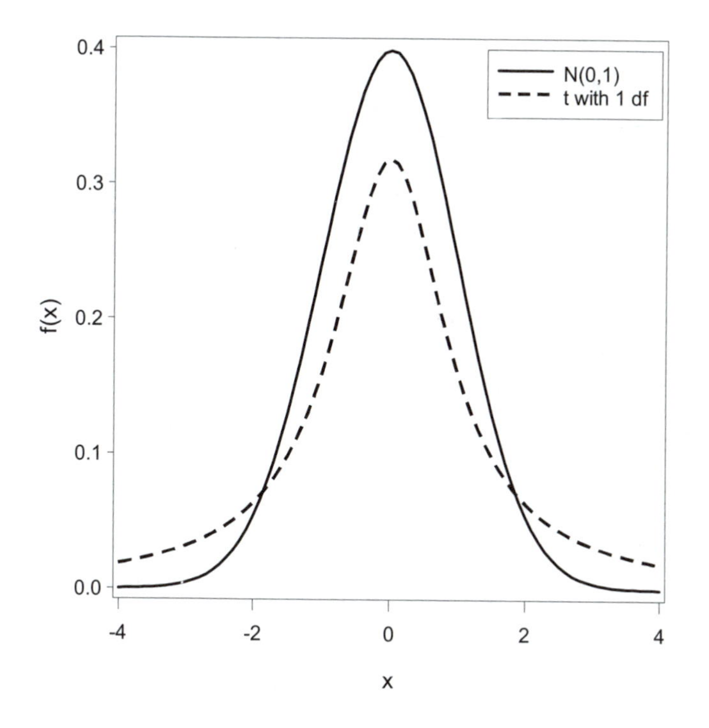

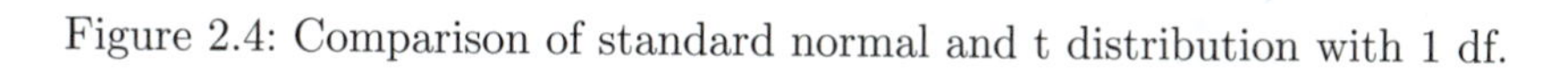
Figure 2.4: Comparison of standard normal and t distribution with 1 df.

Chapter 3

Common statistical procedures

This chapter describes how to generate univariate summary statistics for continuous variables (such as means, variances, and quantiles), display and analyze frequency tables and cross-tabulations for categorical variables, as well as carry out a variety of one- and two-sample procedures. Many of the statistics, tests, and procedures described in this chapter can also be accessed through one of the point-and-click interfaces (1.9).

3.1 Summary statistics

3.1.1 Means and other summary statistics

Example: See 3.6.1

```
proc means data=ds keyword1 ... keywordn;
  var x1 ... xk;
run;
```

or

```
proc summary data=ds;
  var x1 ... xk;
  output out=newds keyword1= keyword2(x2)=newname
    keyword3(x3 x4)=newnamea newnameb;
run;

proc print data=newds;
run;
```

or

```
proc univariate data=ds;
  var x1 ... xk;
run;
```

The `univariate` procedure generates a number of statistics by default, including the mean, standard deviation, skewness, and kurtosis. The `means` and `summary` procedures accept a number of `keywords`, including `mean`, `median`, `var`, `stdev`, `min`, `max`, `sum`. These procedures are identical except that `proc summary` produces no printed output, only an output dataset, while `proc means` can produce both printed output and a dataset. The `output` statement syntax is `keyword=` in which case the summary statistic shares the name of the variable summarized, `keyword(varname)=newname` in which case the summary statistic takes the new name, or `keyword(varname1 ... varnamek)=newname1 ... newnamek` which allows the naming of many summary statistic variables at once. These options become valuable especially when summarizing within subgroups (3.1.2). The `maxdec` option to the `proc means` statement controls the number of decimal places printed.

3.1.2 Means by group

Example: See 3.6.4 and 2.13.4

```
proc sort data=ds;
  by y;
run;

proc means data=ds;
  by y;
  var x;
run;
```

or

```
proc sort data=ds;
  by y;
run;

proc summary data=ds;
  by y;
  output out=newds mean=;
  var x;
run;

proc print data=newds;
run;
```

The summary statistics for each by group are included in any printed output and in any datasets created by the procedure. See Section 3.1.1 for a discussion of output statement syntax.

3.1.3 Trimmed mean

```
proc univariate data=ds trimmed=frac;
  var x;
run;
```

The parameter frac is the proportion of observations above and below the mean to exclude, or a number (greater than 1) in which case number observations will be excluded. Multiple variables may be specified. This statistic can be saved into a dataset using ODS (see 1.7).

3.1.4 Five-number summary

Example: See 3.6.1

The five-number summary (minimum, 25th percentile, median, 75th percentile, maximum) is a useful summary of the distribution of observed values.

```
proc means data=ds mean min q1 median q3 max;
  var x1 ... xk;
run;
```

3.1.5 Quantiles

Example: See 3.6.1

```
proc univariate data=ds;
  var x1 ... xk;
  output out=newds pctlpts=2.5, 95 to 97.5 by 1.25
    pctlpre=p pctlnames=2_5 95 96_125 97_5;
run;
```

This creates a new dataset with the 2.5, 95, 96.25, 97.5 values stored in variables named p2_5, p95, p96_125, and p97_5. The first, 5th, 10th, 25th, 50th, 75th, 90th, 95th, and 99th can be obtained more directly from proc means, proc summary, and proc univariate.

Details and options regarding calculation of quantiles in proc univariate can be found in SAS online help: Contents; SAS Products; SAS Procedures; UNIVARIATE; Calculating Percentiles.

3.1.6 Centering, normalizing, and scaling

```
proc standard data=ds out=ds2 mean=0 std=1;
  var x1 ... xk;
run;
```

The output dataset named in the `out` option contains all of the data from the original dataset, with the standardized version of each variable named in the `var` statement stored in place of the original. Either the `mean` or the `std` option may be omitted.

3.1.7 Mean and 95% confidence interval

```
proc means data=ds lclm mean uclm;
  var x;
run;
```

Calculated statistics can be saved using an `output` statement or using `proc summary` as in 3.1.1 or using `ODS`.

3.1.8 Bootstrapping a sample statistic

Bootstrapping is a powerful and elegant approach to estimation of sample statistics that can be implemented even in many situations where asymptotic results are difficult to find or otherwise unsatisfactory [12]. Bootstrapping proceeds using three steps: first, resample the dataset (with replacement) a specified number of times (typically on the order of 10,000), calculate the desired statistic from each resampled dataset, then use the distribution of the resampled statistics to estimate the standard error of the statistic (normal approximation method), or construct a confidence interval using quantiles of that distribution (percentile method).

As an example, we consider estimating the standard error and 95% confidence interval for the coefficient of variation (COV), defined as σ/μ, for a random variable X. The user must provide code to calculate the statistic of interest; this must be done in a macro.

```
/* download "jackboot.sas" from
  http://support.sas.com/kb/24/982.html */
%include 'c:/sasmacros/jackboot.sas';

/* create macro that generates the desired statistic, in this
   case the coefficient of variation, just once, from the
   observed data.  This macro must be named %analyze  */
%macro analyze(data=, out=);
proc summary data=&data;
  var x;
  output out=&out (drop=_freq_ _type_) cv=cv_x;
run;
%mend;

/* run the boot macro */
%boot(data=ds, samples=1000);
```

The `%include` statement is equivalent to typing the contents of the included file into the program. The `%boot` macro requires an existing `%analyze` macro, which must generate an output dataset; bootstrap results for all variables in this output dataset are calculated. The `drop` dataset option removes some character variables from this output dataset so that statistics are not reported on them. See Section 1.8 for more information on SAS macros.

3.1.9 Proportion and 95% confidence interval

Example: See 7.2.2

```
proc freq data=ds;
  tables x / binomial;
run;
```

The binomial option requests the exact Clopper–Pearson confidence interval based on the F distribution [5], an approximate confidence interval, and a test that the probability of the first level of the variable = 0.5. If x has more than two levels, the probability estimated and tested is the probability of the first level vs. all the others combined. Additional confidence intervals are available as options to the `binomial` option.

3.1.10 Tests of normality

```
proc univariate data=ds normal;
  var x;
run;
```

The `normal` option generates an extra section of output containing four tests for normality.

3.2 Bivariate statistics

3.2.1 Epidemiologic statistics

Example: See 3.6.3

```
proc freq data=ds;
  tables x*y / relrisk;
run;
```

The `freq` procedure will also generate one-way tables, as in 3.1.9.

3.2.2 Test characteristics

The sensitivity of a test is defined as the probability that someone with the disease (D=1) tests positive (T=1), while the specificity is the probability that someone without the disease (D=0) tests negative (T=0). For a dichotomous screening measure, the sensitivity and specificity can be defined as $P(D = 1, T = 1)/P(D = 1)$ and $P(D = 0, T = 0)/P(D = 0)$, respectively. (See also receiver operating character curves, 6.1.17.)

```
proc freq data=ds;
  tables d*t / out=newds;
run;

proc means data=newds nway;
  by d;
  var count;
  output out=newds2 sum=sumdlev;
run;

data newds3;
  merge newds newds2;
  by d;
  retain sens spec;
  if D eq 1 and T=1 then sens=count/sumdlev;
  if D eq 0 and T=0 then spec=count/sumdlev;
  if sens ge 0 and spec ge 0;
run;
```

The above code creates a dataset with a single line containing the sensitivity, specificity, and other data, given a test positive indicator `t` and disease indicator `d`. Sensitivity and specificity across all unique cut-points of a continuous measure `T` can be calculated as follows.

```
proc summary data=ds;
  var d;
  output out=sumdisease sum=totaldisease n=totalobs;
run;

proc sort data=ds; by descending t; run;

data ds2;
  set ds;
  if _n_ eq 1 then set sumdisease;
  retain sumdplus 0 sumdminus 0;
  sumdplus = sumdplus + d;
  sumdminus = sumdminus + (d eq 0);
  sens = sumdplus/totaldisease;
  one_m_spec = sumdminus/(totalobs - totaldisease);
run;
```

In the preceding code, `proc summary` (Section 3.1.1) is used to find the total number with the disease and in the dataset, and to save this data in a dataset named `sumdisease`. The data is then sorted in descending order of the test score `t`. In the final step, the disease and total number of observations are read in and the current number of true positives and negatives accrued as the value of `t` decreases. The conditional use of the `set` statement allows the summary values for disease and subjects to be included for each line of the output dataset; the `retain` statement allows values to be kept across entries in the dataset and optionally allows the initial value to be set. The final dataset contains the sensitivity `sens` and 1 minus the specificity `one_m_spec`. This approach would be more complicated if tied values of the test score were possible.

3.2.3 Correlation

Example: See 3.6.2 and 6.6.6

```
proc corr data=ds;
  var x1 ... xk;
run;
```

Specifying `spearman` or `kendall` as an option to `proc corr` generates the Spearman or Kendall correlation coefficients, respectively. The `with` statement can be used to generate correlations only between the `var` and `with` variables, as in 3.6.2, rather than among all the `var` variables. This can save space as it avoids replicating correlations above and below the diagonal of the correlation matrix.

3.2.4 Kappa (agreement)

```
proc freq data=ds;
  tables x * y / agree;
run;
```

The agree statement produces κ and weighted κ and their asymptotic standard errors and confidence interval, as well as McNemar's test for 2×2 tables, and Bowker's test of symmetry for tables with more than two levels [3].

3.3 Contingency tables

3.3.1 Display cross-classification table

Example: See 3.6.3

Contingency tables show the group membership across categorical (grouping) variables. They are also known as cross-classification tables, cross-tabulations, and two-way tables.

```
proc freq data=ds;
  tables x * y;
run;
```

3.3.2 Pearson's χ^2

Example: See 3.6.3

```
proc freq data=ds;
  tables x * y / chisq;
run;
```

For 2×2 tables the output includes both unadjusted and continuity-corrected tests.

3.3.3 Cochran–Mantel–Haenszel test

The Cochran–Mantel–Haenszel test gives an assessment of the relationship between X_2 and X_3, stratified by (or controlling for) X_1. The analysis provides a way to adjust for the possible confounding effects of X_1 without having to estimate parameters for them.

```
proc freq data=ds;
  tables x1 * x2 * x3 / cmh;
run;
```

The `cmh` option produces Cochran–Mantel–Haenszel statistics and, when both X_2 and X_3 have two values, it generates estimates of the common odds ratio, common relative risks, and the Breslow–Day test for homogeneity of the odds ratios. More complex models can be fit using the generalized linear model methodology described in Chapter 5.

3.3.4 Fisher's exact test

Example: See 3.6.3

```
proc freq data=ds;
  tables x * y / exact;
run;
```

or

```
proc freq data=ds;
  tables x * y;
  exact fisher / mc n=bnum;
run;
```

The former requests only the exact p-value; the latter generates a Monte Carlo p-value, an asymptotically equivalent test based on `bnum` random tables simulated using the observed margins.

3.3.5 McNemar's test

McNemar's test tests the null hypothesis that the proportions are equal across matched pairs, for example, when two raters assess a population.

```
proc freq data=ds;
  tables x * y / agree;
run;
```

3.4 Two sample tests for continuous variables

3.4.1 Student's t-test

Example: See 3.6.4

```
proc ttest data=ds;
  class x;
  var y;
run;
```

The variable X takes on two values. The output contains both equal and unequal-variance t-tests, as well as a test of the null hypothesis of equal variance.

3.4.2 Nonparametric tests

Example: See 3.6.4

```
proc npar1way data=ds wilcoxon edf median;
  class y;
  var x;
run;
```

Many tests can be requested as options to the `proc npar1way` statement. Here we show a Wilcoxon test, a Kolmogorov–Smirnov test, and a median test respectively. Exact tests can be generated by using an `exact` statement with these names, e.g., the `exact median` statement will generate the exact median test.

3.4.3 Permutation test

Example: See 3.6.4

```
proc npar1way data=ds;
  class y;
  var x;
  exact scores=data;
run;
```

or

```
proc npar1way data=ds;
  class y;
  var x;
  exact scores=data / mc n=bnum;
run;
```

Any test described in 3.4.2 can be named in place of `scores=data` to get an exact test based on those statistics. The `mc` option generates an empirical p-value (asymptotically equivalent to the exact p-value) based on `bnum` Monte Carlo replicates.

3.4.4 Logrank test

Example: See 3.6.5

See also 6.1.18 (Kaplan–Meier plot) and 5.3.1 (Cox proportional hazards model)

```
proc phreg data=ds;
  model timevar*cens(0) = x;
run;
```

or

```
proc lifetest data=ds;
  time timevar*cens(0);
  strata x;
run;
```

If `cens` is equal to 0, then `proc phreg` and `proc lifetest` treat *time* as the time of censoring, otherwise the time of the event. The default output from `proc lifetest` includes the logrank and Wilcoxon tests. Other tests, corresponding to different weight functions, can be produced with the `test` option to the `strata` statement. These include `test=fleming(`ρ_1, ρ_2`)`, a superset of the G-rho family of Fleming and Harrington [15], which simplifies to the G-rho family when $\rho_2 = 0$.

3.5 Further resources

Comprehensive introductions to using SAS to fit common statistical models can be found in Cody and Smith [4] and Delwiche and Slaughter [9]. Efron and Tibshirani [12] provide a comprehensive overview of bootstrapping. A readable introduction to permutation-based inference can be found in Good [17]. Collett [6] provides an accessible introduction to survival analysis.

3.6 HELP examples

To help illustrate the tools presented in this chapter, we apply many of the entries to the HELP data. SAS code can be downloaded from `http://www.math.smith.edu/sas/examples`.

3.6.1 Summary statistics and exploratory data analysis

We begin by reading the dataset.

```
filename myurl
  url 'http://www.math.smith.edu/sas/datasets/help.csv'
    lrecl=704;
proc import datafile=myurl out=ds dbms=dlm;
  delimiter=',';
  getnames=yes;
run;
```

The `lrecl` statement is needed due to the long logical lines in the csv file.

A first step would be to examine some univariate statistics (3.1.1) for the baseline CESD (Center for Epidemiologic Studies–Depression) score.

```
options ls=64;  * narrow output to stay in grey box;
proc means data=ds maxdec=2 min p5 q1 median q3 p95 max mean
  std range;
  var cesd;
run;

The MEANS Procedure

                  Analysis Variable : cesd

                                          Lower
     Minimum         5th Pctl          Quartile           Median
-------------------------------------------------------------
        1.00            10.00             25.00            34.00
-------------------------------------------------------------

                  Analysis Variable : cesd

       Upper
    Quartile        95th Pctl           Maximum             Mean
-------------------------------------------------------------
       41.00            53.00             60.00            32.85
-------------------------------------------------------------

  Analysis Variable : cesd

     Std Dev            Range
-----------------------------
       12.51            59.00
-----------------------------
```

We can also generate desired centiles. Here, we find the deciles (3.1.5).

```
ods select none;
proc univariate data=ds;
  var cesd;
  output out=deciles pctlpts= 0 to 100 by 10 pctlpre=p_;
run;
ods select all;

options ls=64;
proc print data=deciles;
run;

Obs p_0 p_10 p_20 p_30 p_40 p_50 p_60 p_70 p_80 p_90 p_100

 1   1   15   22   27   30   34   37   40   44   49    60
```

Graphics can allow us to easily review the whole distribution of the data. Here we generate a histogram (6.1.1) of CESD, overlaid with its empirical PDF (6.1.15) and the closest-fitting normal distribution (see Figure 3.1). The other results of `proc univariate` have been suppressed by `selecting` only the graphics output using an `ods select` statement.

```
ods select univar;
proc univariate data=ds;
  var cesd;
  histogram cesd / normal (color=black l=1)
    kernel(color=black l=21) cfill=greyCC;
run; quit;
ods select all;
```

3.6.2 Bivariate relationships

We can calculate the correlation (3.2.3) between CESD and MCS and PCS (mental and physical component scores). First, we show the default correlation matrix.

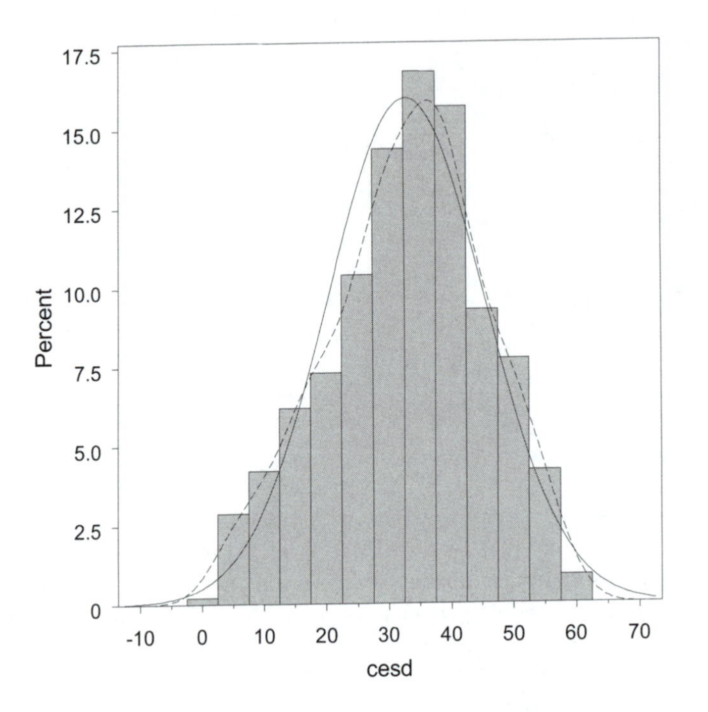

Figure 3.1: Density plot of depressive symptom scores (CESD) plus superimposed histogram and normal distribution.

```
ods select pearsoncorr;
proc corr data=ds;
  var cesd mcs pcs;
run;

The CORR Procedure

              cesd           mcs           pcs

cesd       1.00000      -0.68192      -0.29270
                          <.0001        <.0001

mcs       -0.68192       1.00000       0.11046
            <.0001                      0.0187

pcs       -0.29270       0.11046       1.00000
            <.0001        0.0187
```

The p-value assessing the null hypothesis that the correlation = 0 is printed below each correlation.

To save space, we can just print a subset of the correlations.

```
ods select pearsoncorr;
proc corr data=ds;
  var mcs pcs;
  with cesd;
run;

The CORR Procedure

                   mcs            pcs


cesd          -0.68192       -0.29270
                <.0001         <.0001
```

Figure 3.2 displays a scatterplot (6.1.8) of CESD and MCS, for the female subjects. The plotting character (6.2.2) is the initial letter of the primary substance (Alcohol, Cocaine, or Heroin).

```
symbol1 font=swiss v='A' h=.7 c=black;
symbol2 font=swiss v='C' h=.7 c=black;
symbol3 font=swiss v='H' h=.7 c=black;
proc gplot data=ds;
  where female=1;
  plot mcs*cesd=substance;
run; quit;
```

3.6.3 Contingency tables

Here we display the cross-classification (contingency) table (3.3.1) of homeless at baseline and gender, calculate the observed odds ratio (OR) (3.2.1), and assess association using the Pearson χ^2 test (3.3.2) and Fisher's exact test (3.3.4). This can be done with one call to `proc freq`.

```
proc freq data=ds;
  tables homeless*female / chisq exact relrisk;
run; quit;
```

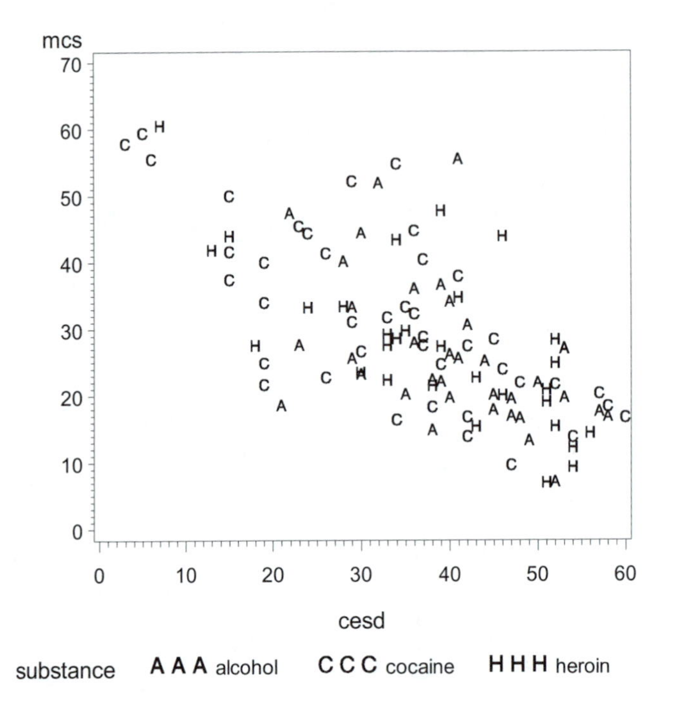

Figure 3.2: Scatterplot of CESD and MCS for women with primary substance shown as the plot symbol.

```
The FREQ Procedure

Table of homeless by female

homeless     female

Frequency|
Percent  |
Row Pct  |
Col Pct  |       0|       1|  Total
---------+--------+--------+
       0 |    177 |     67 |    244
         |  39.07 |  14.79 |  53.86
         |  72.54 |  27.46 |
         |  51.16 |  62.62 |
---------+--------+--------+
       1 |    169 |     40 |    209
         |  37.31 |   8.83 |  46.14
         |  80.86 |  19.14 |
         |  48.84 |  37.38 |
---------+--------+--------+
Total         346      107      453
            76.38    23.62   100.00
```

Within each cell is the number of observations, and the percent of the total, the row, and the column which they make up. Outside the table are the row and column total numbers and their percent of the whole. On the lower right is the total number of observations included in the table.

```
Statistics for Table of homeless by female

Statistic                     DF       Value      Prob
------------------------------------------------------
Chi-Square                     1      4.3196    0.0377
Likelihood Ratio Chi-Square    1      4.3654    0.0367
Continuity Adj. Chi-Square     1      3.8708    0.0491
Mantel-Haenszel Chi-Square     1      4.3101    0.0379
Phi Coefficient                      -0.0977
Contingency Coefficient               0.0972
Cramer's V                           -0.0977
```

Fisher's exact test is provided by default with 2×2 tables, so the `exact` statement is not actually needed here. The exact test result is shown.

```
Statistics for Table of homeless by female

Cell (1,1) Frequency (F)       177
Left-sided Pr <= F          0.0242
Right-sided Pr >= F         0.9861

Table Probability (P)       0.0102
Two-sided Pr <= P           0.0456
```

The following table is requested by the `relrisk` option.

```
Statistics for Table of homeless by female

           Estimates of the Relative Risk (Row1/Row2)

Type of Study                  Value      95% Confidence Limits
-----------------------------------------------------------------
Case-Control (Odds Ratio)     0.6253       0.4008          0.9755
Cohort (Col1 Risk)            0.8971       0.8105          0.9930
Cohort (Col2 Risk)            1.4347       1.0158          2.0265
```

The results suggest that there is a statistically significant association between gender and homelessness, and that men are more likely to be homeless than women.

3.6.4 Two sample tests of continuous variables

We can assess gender differences in baseline age using a t-test (3.4.1) and non-parametric procedures.

```
options ls=64;  /* narrows output to stay in the grey box */

proc ttest data=ds;
  class female;
  var age;
run;
```

```
Variable:  age

female              N        Mean     Std Dev     Std Err

0                 346     35.4682      7.7501      0.4166
1                 107     36.2523      7.5849      0.7333
Diff (1-2)                -0.7841      7.7116      0.8530

female            Minimum     Maximum

0                 19.0000     60.0000
1                 21.0000     58.0000
Diff (1-2)
```

```
Variable:  age

female      Method                Mean      95% CL Mean      Std Dev

0                              35.4682   34.6487  36.2877     7.7501
1                              36.2523   34.7986  37.7061     7.5849
Diff (1-2)  Pooled             -0.7841   -2.4605   0.8923     7.7116
Diff (1-2)  Satterthwaite      -0.7841   -2.4483   0.8800

female      Method              95% CL Std Dev

0                               7.2125    8.3750
1                               6.6868    8.7637
Diff (1-2)  Pooled              7.2395    8.2500
Diff (1-2)  Satterthwaite
```

```
Variable:  age

Method           Variances        DF    t Value    Pr > |t|

Pooled           Equal           451      -0.92      0.3585
Satterthwaite    Unequal      179.74      -0.93      0.3537
```

```
Variable:  age

              Equality of Variances

Method      Num DF    Den DF    F Value    Pr > F

Folded F       345       106       1.04    0.8062
```

The output of the `proc ttest` output is particularly awkward, and the formatting for this book makes it worse. The output shows that there are 346 men with a mean age of 35.5 years, with a standard deviation of 7.75 years, and that the men range in age between 19 and 60 years. In the next block of output, the means per group are repeated, along with 95% confidence interval (CI) for the means, the standard deviations are also repeated, along with 95% CI for the standard deviations. The differences between the means is shown in the `Diff` rows, where the difference and CI for the difference is shown. Assuming equal variances for the groups allows a single estimated standard deviation. This is shown, along with its CI in the `Pooled` rows. The third box shows the result of the test. The final row of output shows a test of the null hypothesis of equal variances.

A permutation test can be run and used to generate a Monte Carlo p-value (3.4.3).

```
ods select datascoresmc;
proc npar1way data=ds;
  class female;
  var age;
  exact scores=data / mc n=9999 alpha=.05;
run;
ods exclude none;

One-Sided Pr >= S
Estimate                           0.1841
95% Lower Conf Limit               0.1765
95% Upper Conf Limit               0.1917

Two-Sided Pr >= |S - Mean|
Estimate                           0.3615
95% Lower Conf Limit               0.3521
95% Upper Conf Limit               0.3710

Number of Samples                    9999
Initial Seed                    600843001
```

The Monte Carlo test is an exact test, and thus the displayed confidence limits are not appropriate [11].

Both the Wilcoxon test and Kolmogorov–Smirnov test (3.4.2) can be generated by a single `proc npar1way`. We will include the D statistic from the Kolmogorov–Smirnov test and the associated p-value in a figure title; to make that possible, we will use `ODS` to create a dataset containing these values.

```
ods output kolsmir2stats=age_female_ks_stats;
ods select wilcoxontest kolsmir2stats;
proc npar1way data=ds wilcoxon edf;
  class female;
  var age;
run;
ods select all;
```

```
Statistic                  25288.5000

Normal Approximation
Z                              0.8449
One-Sided Pr >  Z              0.1991
Two-Sided Pr > |Z|             0.3981

t Approximation
One-Sided Pr >  Z              0.1993
Two-Sided Pr > |Z|             0.3986

Z includes a continuity correction of 0.5.
```

```
KS    0.026755     D           0.062990
KSa   0.569442     Pr > KSa    0.9020
```

The result from each test is that the null hypothesis of identical distributions cannot be rejected at the 5% level.

We can also plot estimated density functions (6.1.15) for age for both groups, and shade some areas (6.2.12) to emphasize how they differ (Figure 3.3). The `univariate` procedure with a `by` statement will generate density estimates for each group, but not overplot them. To overplot, we first generate the density estimates for each gender using `proc kde` (6.1.15) (suppressing all printed output).

```
proc sort data=ds;
  by female;
run;

ods select none;
proc kde data=ds;
  by female;
  univar age / out=kdeout;
run;
ods select all;
```

Next, we will review the `proc npar1way` output we saved as a dataset.

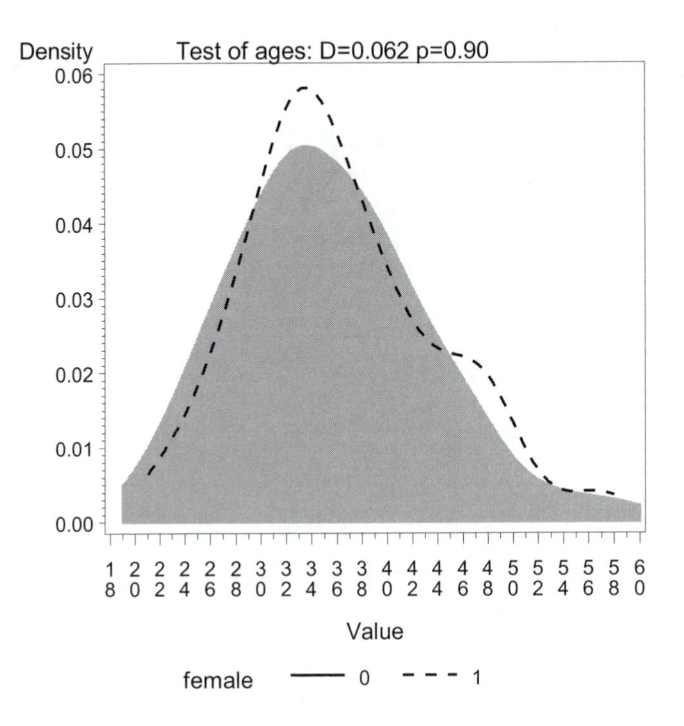

Figure 3.3: Density plot of age by gender.

```
options ls=64;
proc print data=age_female_ks_stats; run;

Obs    Variable     Name1     Label1    cValue1           nValue1

 1       age        _KS_       KS       0.026755         0.026755
 2       age        _KSA_      KSa      0.569442         0.569442

Obs    Name2      Label2      cValue2            nValue2

 1     _D_        D           0.062990           0.062990
 2     P_KSA      Pr > KSa    0.9020             0.901979
```

Running `proc contents` (2.3.1, results not shown) reveals that the variable names prepended with 'c' are character variables. To get these values into a figure title, we create SAS Macro variables (1.8.2) using the `call symput` function.

```
data _null_;
set age_female_ks_stats;
  if label2 eq 'D' then
    call symput('dvalue', substr(cvalue2, 1, 5));
    /* This makes a macro variable (which is saved outside any
            dataset) from a value in a dataset */
  if label2 eq 'Pr > KSa' then
    call symput('pvalue', substr(cvalue2, 1, 4));
run;
```

Finally, we construct the plot using `proc gplot` for the data with a `title` statement to include the Kolmogorov–Smirnov test results.

```
symbol1 i=j w=5 l=1 v=none c=black;
symbol2 i=j w=5 l=2 v=none c=black;
title "Test of ages: D=&dvalue p=&pvalue";
pattern1 color=grayBB;
proc gplot data=kdeout;
  plot density*value = female /
    legend areas=1 haxis=18 to 60 by 2;
run; quit;
```

In this code, the `areas` option to the `plot` statement makes SAS fill in the area under the first curve, while the `pattern` statement describes what color to fill in with. The plot confirms the results of the tests: the estimated densities look similar.

3.6.5 Survival analysis: Logrank test

The logrank test (3.4.4) can be used to compare estimated survival curves between groups in the presence of censoring. Here we compare randomization groups with respect to `dayslink`. A value of 0 for `linkstatus` indicates that the observation was censored, not observed, at the time recorded in `dayslink`.

```
ods select homtests;
proc lifetest data=ds;
  time dayslink*linkstatus(0);
  strata treat;
run;
ods select all;

       Test of Equality over Strata

                                          Pr >
Test         Chi-Square       DF    Chi-Square

Log-Rank        84.7878        1        <.0001
Wilcoxon        87.0714        1        <.0001
-2Log(LR)      107.2920        1        <.0001
```

Chapter 4

Linear regression and ANOVA

Regression and analysis of variance form the basis of many investigations. In this chapter we describe how to undertake many common tasks in linear regression (broadly defined), while Chapter 5 discusses many generalizations, including other types of outcome variables, longitudinal and clustered analysis, and survival methods.

Many procedures can perform linear regression, as it constitutes a special case of which many models are generalizations. We present detailed descriptions for `proc reg` and `proc glm`, as these offer the most flexibility and best output options tailored to linear regression in particular. While analysis of variance (ANOVA) can be viewed as a special case of linear regression, separate routines are available (e.g., `proc anova`) to perform it. In addition, `proc mixed` is also useful for some calculations. We address these additional procedures only with respect to output that is difficult to obtain through the standard linear regression tools. Linear regression and ANOVA can also be accessed through one of the point-and-click interfaces (1.9).

4.1 Model fitting

4.1.1 Linear regression

Example: See 4.7.3

```
proc glm data=ds;
  model y = x1 ... xk;
run;
```

or

```
proc reg data=ds;
  model y = x1 ... xk;
run;
```

Both `proc glm` and `proc reg` support linear regression models, while `proc reg` provides more regression diagnostics. The `glm` procedure more easily allows categorical covariates.

4.1.2 Linear regression with categorical covariates

Example: See 4.7.3

See also 4.1.3 (parameterization of categorical covariates)

```
proc glm data=ds;
  class x1;
  model y = x1 x2 ... xk;
run;
```

The `class` statement specifies covariates that should be treated as categorical. The `glm` procedure uses reference cell coding; the reference category can be controlled using the `order` option to the `proc glm` statement, as in 5.6.11.

4.1.3 Parameterization of categorical covariates

Example: See 4.7.6

Some procedures accept a `class` statement to declare that a covariate should be treated as categorical. Of the model-fitting procedures mentioned in this book, the following procedures will not accept a class statement, as of SAS 9.2: `arima`, `catmod`, `countreg`, `factor`, `freq`, `kde`, `lifetest`, `nlin`, `nlmixed`, `reg`, `surveyfreq`, and `varclus`. For these procedures, indicator (or "dummy") variables must be created in a data step.

The following procedures accept a class statement which applies reference cell or indicator variable coding to the listed variables: `proc anova`, `candisc`, `discrim`, `gam`, `glimmix`, `glm`, `mi`, `mianalyze`, `mixed`, `quantreg`, `robustreg`, `stepdisc`, and `surveyreg`. The value used as the referent can often be controlled, usually as an `order` option to the controlling `proc`, as in 5.6.11. For these procedures, other parameterizations must be coded in a data step. The following procedures accept multiple parameterizations, using the syntax shown below for `proc logistic`: `proc genmod` (defaults to reference cell coding), `proc glmselect` (defaults to reference cell coding), `proc logistic` (defaults to effect coding), `proc phreg` (defaults to reference cell coding), and `proc surveylogistic` (defaults to effect coding). Finally, `proc univariate` accepts a `class` statement, but uses it only to generate results for each level of the named variables, so that no coding is implied.

```
proc logistic data=ds;
  class x1 (param=paramtype) x2 (param=paramtype);
  ...
run;
```

or

```
proc logistic data=ds;
  class x1 x2 / param=paramtype;
  ...
run;
```

Available `paramtypes` include: (1) `orthpoly` (orthogonal polynomials); (2) `effect` (the default for `proc logistic` and `proc surveylogistic`); and (3) `ref` or `glm`, (reference cell or dummy coding). In addition, if the same parameterization is desired for all of the categorical variables in the model, it can be added in a statement such as the second example. In this case, `param=glm` can be used to emulate the parameterization found in the other procedures which accept `class` statements; this is the default for `proc genmod` and `proc phreg`.

4.1.4 Linear regression with no intercept

```
proc glm data=ds;
  model y = x1 ... xk / noint;
run;
```

The `noint` option works with the `model` statements in many procedures.

4.1.5 Linear regression with interactions

Example: See 4.7.3

```
proc glm data=ds;
  model y = x1 x2 x1*x2 x3 ... xk;
run;
```

or

```
proc glm data=ds;
  model y = x1|x2 x3 ... xk;
run;
```

The `|` operator includes the product and all lower order terms, while the `*` operator includes only the specified interaction. So, for example, `model y = x1|x2|x3` and `model y = x1 x2 x3 x1*x2 x1*x3 x2*x3 x1*x2*x3` are equivalent statements. The syntax above also works with any covariates designated as categorical using the class statement (4.1.2). The model statement for many procedures accepts this syntax.

4.1.6 Linear models stratified by each value of a grouping variable

Example: See 4.7.5

It is easy to fit models stratified by each value of a grouping variable (see also subsetting, 2.5.1).

```
proc sort data=ds;
  by z;
run;

ods output parameterestimates=params;
proc reg data=ds;
  by z;
  model y = x1 ... xk;
run;
```

Note that if the by variable has many distinct values, output may be voluminous. A single dataset containing the parameter estimates from each by group (1.6.2) can be created by issuing an ods output parameterestimates=ds statement before the proc reg statement.

4.1.7 One-way analysis of variance

Example: See 4.7.6

```
proc glm data=ds;
  class x;
  model y = x / solution;
run;
```

The solution option to the model statement requests that the parameter estimates be displayed. Other procedures which fit ANOVA models include proc anova and proc mixed.

4.1.8 Two-way (or more) analysis of variance

Example: See 4.7.6

Interactions can be specified using the syntax introduced in Section 4.1.5 (see also interaction plots, Section 6.1.11).

```
proc glm data=ds;
  class x1 x2;
  model y = x1 x2;
run;
```

Other procedures which fit ANOVA models include `proc anova` and `proc mixed`.

4.2 Model comparison and selection

4.2.1 Compare two models

Example: See 4.7.6

In general, most procedures fit a single model. Comparisons between models must be constructed by hand. An exception is "leave-one-out" models, in which a model identical to the one fit is considered, except that a single predictor is to be omitted. In this case, SAS offers "Type III" sums of squares tests, which can be printed by default or requested in many modeling procedures. These Wald tests and likelihood ratio tests are identical in many settings, though they differ in general. In cases in which they differ, likelihood ratio tests are to be preferred.

4.2.2 Log-likelihood

See also 4.2.3 (AIC) *Example:* See 4.7.6

```
proc mixed data=ds;
  model y = x1 ... xk;
run;
```

Log-likelihood values are produced by various procedures, but the syntax to generate them can be idiosyncratic. The `mixed` procedure fits a superset of models available in `proc glm`, and can be used to generate this quantity.

4.2.3 Akaike Information Criterion (AIC)

See also 4.2.2 (log-likelihood) *Example:* See 4.7.6

```
proc reg data=ds stats=aic;
  model y = x1 ... xk;
run;
```

AIC values are available in various procedures, but the syntax to generate them can be idiosyncratic.

4.2.4 Bayesian Information Criterion (BIC)

See also 4.2.3 (AIC)

```
proc mixed data=ds;
  model y = x1 ... xk;
run;
```

BIC values are presented by default in `proc mixed`.

4.3 Tests, contrasts, and linear functions of parameters

4.3.1 Joint null hypotheses: Several parameters equal 0

```
proc reg data=ds;
  model ...;
  nametest: test varname1=0, varname2=0;
run;
```

In the above, `nametest` is an arbitrary label which will appear in the output. Multiple `test` statements are permitted.

4.3.2 Joint null hypotheses: Sum of parameters

```
proc reg data=ds;
  model ...;
  nametest: test varname1 + varname2=1;
run;
```

The `test` statement is prefixed with the arbitrary label `nametest` which will appear in the output. Multiple `test` statements are permitted.

4.3.3 Tests of equality of parameters

Example: See 4.7.8

```
proc reg data=ds;
  model ...;
  nametest: test varname1=varname2;
run;
```

The `test` statement is prefixed with the arbitrary label `nametest` which will appear in the output. Multiple `test` statements are permitted.

4.3.4 Multiple comparisons

Example: See 4.7.7

```
proc glm data=ds;
  class x1;
  model y = x1;
  lsmeans x1 / pdiff adjust=tukey;
run;
```

The `pdiff` option requests p-values for the hypotheses involving the pairwise comparison of means. The `adjust` option adjusts these p-values for multiple comparisons. Other options available through `adjust` include `bon` (for Bonferroni), and `dunnett`, among others. SAS `proc mixed` also has an `adjust` option for its `lsmeans` statement. A graphical presentation of significant differences among levels can be obtained with the `lines` option to the `lsmeans` statement, as shown in 4.7.7.

4.3.5 Linear combinations of parameters

Example: See 4.7.8

It is often useful to calculate predicted values for particular covariate values. Here, we calculate the predicted value $E[Y|X_1 = 1, X_2 = 3] = \hat{\beta}_0 + \hat{\beta}_1 + 3\hat{\beta}_2$.

```
proc glm data=ds;
  model y = x1 ... xk;
  estimate 'label' intercept 1 x1 1 x2 3;
run;
```

The `estimate` statement is used to calculate linear combination of parameters (and associated standard errors). The optional quoted text is a label which will be printed with the estimated function.

4.4 Model diagnostics

4.4.1 Predicted values

Example: See 4.7.3

```
proc reg data=ds;
  model ...;
  output out=newds predicted=predicted_varname;
run;
```

or

```
proc glm data=ds;
  model ...;
  output out=newds predicted=predicted_varname;
run;
```

The `output` statement creates a new dataset and specifies variables to be included, of which the predicted values are an example. Others can be found using the online help: Contents; SAS Products; SAS Procedures; REG; Output Statement.

4.4.2 Residuals

Example: See 4.7.3

```
proc glm data=ds;
  model ...;
  output out=newds residual=residual_varname;
run;
```

or

```
proc reg data=ds;
  model ...;
  output out=newds residual=residual_varname;
run;
```

The `output` statement creates a new dataset and specifies variables to be included, of which the residuals are an example. Others can be found using the online help: Contents; SAS Products; SAS Procedures; Proc REG; Output Statement.

4.4.3 Studentized residuals

Example: See 4.7.3

Standardized residuals are calculated by dividing the ordinary residual (observed minus expected, $y_i - \hat{y}_i$) by an estimate of its standard deviation. Studentized residuals are calculated in a similar manner, where the predicted value and the variance of the residual are estimated from the model fit while excluding that observation. In `proc glm` the standardized residual is requested by the `student` option, while the `rstudent` option generates the studentized residual.

```
proc glm data=ds;
  model ...;
  output out=newds student=standardized_resid_varname;
run;
```

or

```
proc reg data=ds;
  model ...;
  output out=newds rstudent=studentized_resid_varname;
run;
```

The `output` statement creates a new dataset and specifies variables to be included, of which the studentized residuals are an example. Both `proc reg` and `proc glm` include both types of residuals. Others can be found using the online help: Contents; SAS Products; SAS Procedures; Proc REG; Output Statement.

4.4.4 Leverage

Example: See 4.7.3

Leverage is defined as the diagonal element of the $(X(X^TX)^{-1}X^T)$ or "hat" matrix.

```
proc glm data=ds;
  model ...;
  output out=newds h=leverage_varname;
run;
```

or

```
proc reg data=ds;
  model ...;
  output out=newds h=leverage_varname;
run;
```

The `output` statement creates a new dataset and specifies variables to be included, of which the leverage values are one example. Others can be found using the online help: Contents; SAS Products; SAS Procedures; Proc REG; Output Statement.

4.4.5 Cook's D

Example: See 4.7.3

Cook's distance (D) is a function of the leverage (see 4.4.4) and the residual. It is used as a measure of the influence of a data point in a regression model.

```
proc glm data=ds;
  model ...;
  output out=newds cookd=cookd_varname;
run;
```

or

```
proc reg data=ds;
  model ...;
  output out=newds cookd=cookd_varname;
run;
```

The `output` statement creates a new dataset and specifies variables to be included, of which the Cook's distance values are an example. Others can be found using the online help: Contents; SAS Products; SAS Procedures; Proc REG; Output Statement.

4.4.6 DFFITS

Example: See 4.7.3

DFFITS are a standardized function of the difference between the predicted value for the observation when it is included in the dataset and when (only) it is excluded from the dataset. They are used as an indicator of the observation's influence.

```
proc reg data=ds;
  model ...;
  output out=newds dffits=dffits_varname;
run;
```

or

```
proc glm data=ds;
  model ...;
  output out=newds dffits=dffits_varname;
run;
```

The `output` statement creates a new dataset and specifies variables to be included, of which the dffits values are an example. Others can be found using the online help: Contents; SAS Products; SAS Procedures; Proc REG; Output Statement.

4.4.7 Diagnostic plots

Example: See 4.7.4

```
proc reg data=ds;
  model ...
  output out=newds predicted=pred_varname residual=resid_varname
    h=leverage_varname cookd=cookd_varname;
run;

proc gplot data=ds;
  plot resid_varname * pred_varname;
  plot resid_varname * leverage_varname;
run;
quit;
```

Q-Q plots of residuals can be generated via proc univariate. The `ods graphics on` statement (1.7.3), issued prior to running the `reg` procedure will produce many diagnostic plots, as will running `ods graphics on` and then `proc glm` with the `plots=diagnostics` option.

4.5 Model parameters and results

4.5.1 Parameter estimates

Example: See 4.7.3

```
ods output parameterestimates=newds;
proc glm data=ds;
  model ... / solution;
run;
```

or

```
proc reg data=ds outest=newds;
  model ...;
run;
```

The `ods output` statement (Section 1.7.1) can be used to save any piece of output as a dataset. The `outest` option is specific to `proc reg`, though many other procedures accept similar syntax.

4.5.2 Standard errors of parameter estimates

See also 4.5.10 (covariance matrix)

```
proc reg data=ds outest=newds;
  model .../ outseb ...;
run;
```

or

```
ods output parameterestimates=newds;
proc glm data=ds;
  model .../ solution;
run;
```

The `ods output` statement (Section 1.7.1) can be used to save any piece of output as a dataset.

4.5.3 Confidence intervals for parameter estimates

```
ods output parameterestimates=newds;
proc glm data=ds;
  model .../ solution clparm;
run;
```

The `ods output` statement (Section 1.7.1) can be used to save any piece of output as a dataset.

4.5.4 Confidence intervals for the mean

These are the lower (and upper) confidence limits for the mean of observations with the given covariate values, as opposed to the prediction limits for individual observations with those values (see 4.5.5).

```
proc glm data=ds;
  model ...;
  output out=newds lclm=lcl_mean_varname;
run;
```

or

```
proc reg data=ds;
  model ...;
  output out=newds lclm=lcl_mean_varname;
run;
```

The `output` statement creates a new dataset and specifies output variables to be included, of which the lower confidence limit values are one example. The upper confidence limits can be generated using the `uclm` option to the `output` statement. Other possibilities can be found using the online help: Contents; SAS Products; SAS Procedures; Proc REG; Output Statement.

4.5.5 Prediction limits

These are the lower (and upper) prediction limits for "new" observations with the covariate values of subjects observed in the dataset, as opposed to confidence limits for the population mean (see 4.5.4).

```
proc glm data=ds;
  model ...;
  output out=newds lcl=lcl_varname;
run;
```

or

```
proc reg data=ds;
  model ...;
  output out=newds lcl=lcl_varname;
run;
```

The `output` statement creates a new dataset and specifies variables to be included, of which the lower prediction limit values are an example. The upper limits can be requested with the `ucl` option to the `output` statement. Other possibilities can be found using the online help: Contents; SAS Products; SAS Procedures; Proc REG; Output Statement.

4.5.6 Plot confidence intervals for the mean

```
symbol1 i=rlclm95 value=none;
proc gplot data=ds;
  plot y * x;
run;
```

The `symbol` statement `i` option (synonym for `interpolation`) contains many useful options for adding features to scatterplots. The `rlclm95` selection requests a regression line plot, with 95% confidence limits for the mean. The `value=none` requests that the observations themselves not be plotted (see also scatterplots, 6.1.8).

4.5.7 Plot prediction limits from a simple linear regression

```
symbol1 i=rlcli95 l=2 value=none;
proc gplot data=ds;
  plot y * x;
run;
```

The symbol statement i (synonym for interpolation) option contains many useful options for adding features to scatterplots. The rlcli95 selection requests a regression line plot, with 95% confidence limits for the values. The value=none requests that the observations not be plotted. The symbol statement "i" option contains many useful features which can be added to scatterplots (see also 6.1.8).

4.5.8 Plot predicted lines for each value of a variable

Here we describe how to generate plots for a variable X_1 versus Y separately for each value of the variable X_2 (see also conditioning plot, 6.1.12).

```
symbol1 i=rl value=none;
symbol2 i=rl value=none;
proc gplot data=ds;
  plot y*x1 = x2;
run;
```

The symbol statement i (synonym for interpolation) option contains many useful options for adding features to scatterplots. The rl selection requests a regression line plot. The value=none requests that the observations not be plotted. The = x2 syntax requests a different symbol statement be applied for each level of x2 (see also scatterplots, 6.1.8).

4.5.9 SSCP matrix

See also 2.9 (matrices)

```
proc reg data=ds;
  model .../ xpx ...;
run;
```

or

```
proc glm data=ds;
  model .../ xpx ...;
run;
```

A dataset containing the information $(X'X)$ matrix can be created using the ODS system with either proc statement or by adding the option outsscp=newds to the proc reg statement.

4.5.10 Covariance matrix

Example: See 4.7.3

See also 2.9 (matrices) and 4.5.2 (standard errors)

```
proc reg data=ds outest=newds covout;
  ...
run;
```

or

```
ods output covb=newds;
proc reg data=ds;
  model ... / covb ...;
run;
```

4.6 Further resources

Accessible guides to linear regression can be found in Littell, Stroup, and Freund [29]. Cook [8] reviews regression diagnostics.

4.7 HELP examples

To help illustrate the tools presented in this chapter, we apply many of the entries to the HELP data. SAS code can be downloaded from http://www.math.smith.edu/sasbook/examples.

We begin by reading in the dataset and keeping only the female subjects.

```
proc import datafile='c:/book/help.dta' out=help_a dbms=dta;
run;

data help;
set help_a;
  if female;
run;
```

4.7.1 Scatterplot with smooth fit

As a first step to help guide fitting a linear regression, we create a scatterplot (6.1.8) displaying the relationship between age and the number of alcoholic drinks consumed in the period before entering detox (variable name: i1), as well as primary substance of abuse (alcohol, cocaine, or heroin).

Figure 4.1 displays a scatterplot of observed values for i1 (along with separate smooth fits by primary substance). To improve legibility, the plotting region is restricted to those with number of drinks between 0 and 40 (see plotting limits, 6.3.7).

```
axis1 order = (0 to 40 by 10) minor=none;
axis2 minor=none;
legend1 label=none value=(h=1.5) shape=symbol(10,1.2)
  down=3 position=(top right inside) frame mode=protect;
symbol1 v=circle i=sm70s c=black l=1 h=1.1 w=5;
symbol2 v=diamond i=sm70s c=black l=33 h=1.1 w=5;
symbol3 v=square i=sm70s c=black l=8 h=1.1 w=5;
proc gplot data=help;
  plot i1*age = substance /
    vaxis=axis1 haxis=axis2 legend=legend1;
run; quit;
```

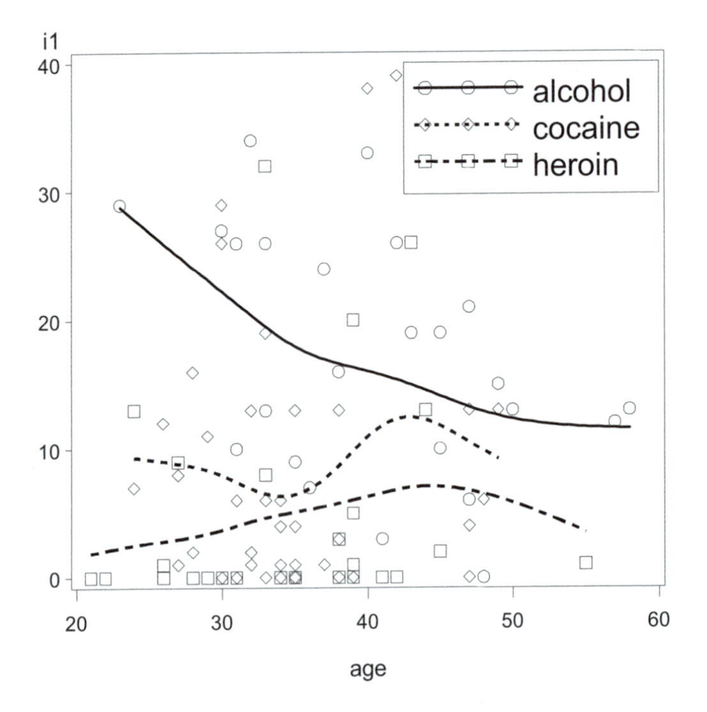

Figure 4.1: Scatterplot of observed values for AGE and I1 (plus smoothers by substance).

Not surprisingly, Figure 4.1 suggests that there is a dramatic effect of primary substance, with alcohol users drinking more than others. There is some indication of an interaction with age. It is important to note that SAS uses only the points displayed (i.e., within the specified axes) when smoothing.

4.7.2 Regression with prediction intervals

We demonstrate plotting confidence limits as well as prediction limits (4.5.6 and 4.5.7) from a linear regression model of `pcs` as a function of `age`. The `symbol` statement options described in 4.5.6, 4.5.7, and 6.2.5 will easily add one or the other of these limits to the plot, but not both at the same time. To get them both, we print the scatterplot twice, with the same data but different limits requested, and use the `overlay` option (6.1.9) to the `plot` statement to make both plots appear together. We do not need to plot the data points twice, so we define the symbol value to be `none` in the second plot.

Figure 4.2 displays the predicted line along with these intervals.

```
symbol1 v=dot h=1.1 w=5 i=rlcli c=black l=1;
symbol2 v=none w=5 i=rlclm c=black l=1;
proc gplot data=help;
  plot (pcs pcs)*age/overlay;
run; quit;
```

4.7.3 Linear regression with interaction

Next we fit a linear regression model (4.1.1) for the number of drinks as a function of age, substance, and their interaction (4.1.5). To assess the need for the interaction, we use the F test from the Type III sums of squares in SAS. To save space, some results of `proc glm` have been suppressed using the `ods select` statement (see 1.7).

```
options ls=64;  /* keep output in gray area */
ods select overallanova modelanova parameterestimates;
proc glm data=help;
class substance;
  model i1 = age substance age * substance / solution;
run; quit;
ods select all;
```

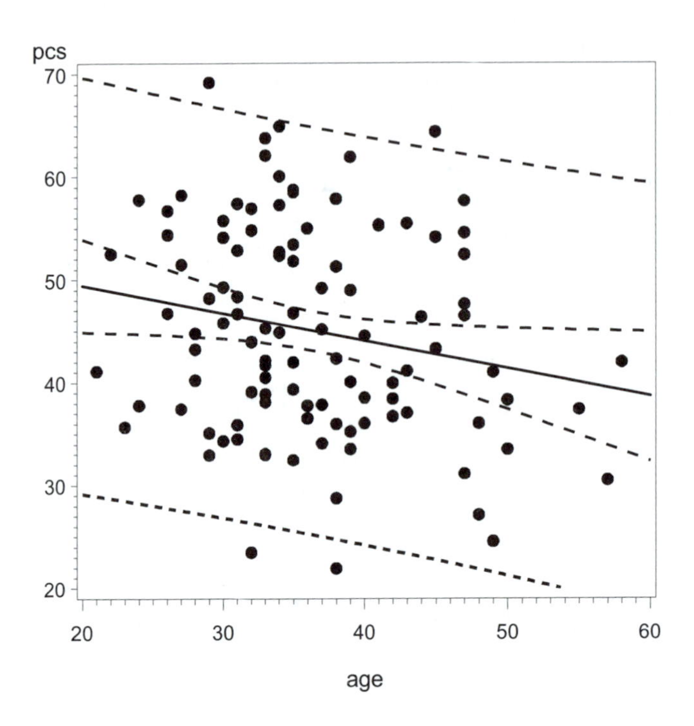

Figure 4.2: Predicted mean, confidence limits for the mean, and prediction limits for new observations.

```
The GLM Procedure

Dependent Variable: I1   i1

                                         Sum of
Source                      DF          Squares     Mean Square

Model                        5      12275.17570      2455.03514

Error                      101      24815.36635       245.69670

Corrected Total            106      37090.54206

Source                  F Value    Pr > F

Model                      9.99    <.0001

Error

Corrected Total
```

```
The GLM Procedure

Dependent Variable: I1   i1

Source                  DF      Type I SS     Mean Square

AGE                      1      384.75504       384.75504
SUBSTANCE                2    10509.56444      5254.78222
AGE*SUBSTANCE            2     1380.85622       690.42811

Source              F Value    Pr > F

AGE                    1.57    0.2137
SUBSTANCE             21.39    <.0001
AGE*SUBSTANCE          2.81    0.0649


Source                  DF    Type III SS     Mean Square

AGE                      1      27.157727       27.157727
SUBSTANCE                2    3318.992822     1659.496411
AGE*SUBSTANCE            2    1380.856222      690.428111

Source              F Value    Pr > F

AGE                    0.11    0.7402
SUBSTANCE              6.75    0.0018
AGE*SUBSTANCE          2.81    0.0649
```

There is some indication of a borderline significant interaction between age and substance group (p=0.065), displayed in the `Type III SS` table, which presents the "leave-one out" analysis. The table shows that the main effect of `age` adds little to a model including `substance` and the interaction between `age` and `substance`. The `Type I SS` table shows the usefulness of each predictor when added to the previous model. Thus a model with just `age` would show a p-value of 0.21.

```
The GLM Procedure

Dependent Variable: I1   i1

                                                  Standard
Parameter                        Estimate            Error    t Value

Intercept                    -7.77045212 B     12.87885672      -0.60
AGE                           0.39337843 B      0.36221749       1.09
SUBSTANCE      alcohol       64.88044165 B     18.48733701       3.51
SUBSTANCE      cocaine       13.02733169 B     19.13852222       0.68
SUBSTANCE      heroin         0.00000000 B       .                .
AGE*SUBSTANCE alcohol        -1.11320795 B      0.49135408      -2.27
AGE*SUBSTANCE cocaine        -0.27758561 B      0.53967749      -0.51
AGE*SUBSTANCE heroin          0.00000000 B       .                .

Parameter                    Pr > |t|

Intercept                     0.5476
AGE                           0.2801
SUBSTANCE      alcohol        0.0007
SUBSTANCE      cocaine        0.4976
SUBSTANCE      heroin          .
AGE*SUBSTANCE alcohol         0.0256
AGE*SUBSTANCE cocaine         0.6081
AGE*SUBSTANCE heroin           .
```

The preceding output is caused by the `solution` option. From this we can see that the the nonsignificant `age` effect in the previous output refers to the effect of age among heroin users. The regression line for alcohol users has an intercept of $64.9 - 7.8 = 57.1$ drinks and a slope of $-1.1 + 0.4 = -0.7$ fewer drinks per year of age.

The `ods output` statement can be used to save any printed result as a SAS dataset. In the following code, all printed output from `proc glm` is suppressed, but the parameter estimates are saved as a SAS dataset, then printed using `proc print`. In addition, various diagnostics are saved via the `output` statement.

```
ods select none;
ods output parameterestimates=helpmodelanova;
proc glm data=help;
  class substance;
  model i1 = age|substance / solution;
  output out=helpout cookd=cookd_ch4 dffits=dffits_ch4
    student=sresids_ch4 residual=resid_ch4
    predicted=pred_ch4 h=lev_ch4;
run; quit;
ods select all;
```

```
options ls=64; /* keep output in gray area */
proc print data=helpmodelanova;
  var parameter estimate stderr tvalue probt;
  format _numeric_ 6.3;
run;
```

```
Obs  Parameter                  Estimate  StdErr  tValue   Probt

 1   Intercept                    -7.770  12.879  -0.603   0.548
 2   AGE                           0.393   0.362   1.086   0.280
 3   SUBSTANCE       alcohol      64.880  18.487   3.509   0.001
 4   SUBSTANCE       cocaine      13.027  19.139   0.681   0.498
 5   SUBSTANCE       heroin        0.000       .       .       .
 6   AGE*SUBSTANCE alcohol        -1.113   0.491  -2.266   0.026
 7   AGE*SUBSTANCE cocaine        -0.278   0.540  -0.514   0.608
 8   AGE*SUBSTANCE heroin          0.000       .       .       .
```

4.7.4 Regression diagnostics

Assessing the model is an important part of any analysis. We begin by examining the residuals (4.4.2). First, we calculate the quantiles of their distribution, then display the smallest residual.

```
options ls=64;  /* keep output in gray area */
proc means data=helpout min q1 median q3 max maxdec=2;
  var resid_ch4;
run;

The MEANS Procedure

                Analysis Variable : resid_ch4

                    Lower                          Upper
     Minimum     Quartile         Median        Quartile
------------------------------------------------------------
      -31.92        -8.31          -4.18            3.69
------------------------------------------------------------

Analysis Variable : resid_ch4

     Maximum
------------
       49.88
------------
```

We could examine the output, then condition to find the value of the residual that is less than −31. Instead the dataset can be sorted so the smallest observation is first and then print one observation.

```
proc sort data=helpout;
  by resid_ch4;
run;

proc print data=helpout (obs=1);
   var id age i1 substance pred_ch4 resid_ch4;
run;

                                                       resid_
Obs      ID     AGE     I1    SUBSTANCE    pred_ch4      ch4

  1     325      35      0     alcohol      31.9160   -31.9160
```

One way to print the largest value is to sort the dataset in the reverse order, then print just the first observation.

```
proc sort data=helpout;
  by descending resid_ch4;
run;

proc print data=helpout (obs=1);
  var id age i1 substance pred_ch4 resid_ch4;
run;

                                                     resid_
Obs    ID    AGE    I1    SUBSTANCE    pred_ch4       ch4

  1     9     50    71     alcohol      21.1185     49.8815
```

Graphical tools are the best way to examine residuals. Figure 4.3 displays the Q-Q plot generated from the saved diagnostics.

Sometimes it is necessary to clear out old graphics settings. This is easiest to do with the `goptions reset=all` statement (6.3.5).

```
goptions reset=all;
```

```
ods select univar;
proc univariate data=helpout;
  qqplot resid_ch4 / normal(mu=est sigma=est);
run;
ods select all;
```

We could use `ods graphics` (1.7.3) to get assorted diagnostic plots, but here we demonstrate a manual approach using the previously saved diagnostics. Figure 4.4 displays the empirical density of the standardized residuals, along with an overlaid normal density. The assumption that the residuals are approximately Gaussian does not appear to be tenable. Further exploration should be undertaken.

```
axis1 label=("Standardized residuals");
ods select "Histogram 1";
proc univariate data=helpout;
  var sresids_ch4;
  histogram sresids_ch4 / normal(mu=est sigma=est color=black)
    kernel(color=black) haxis=axis1;
run;
ods select all;
```

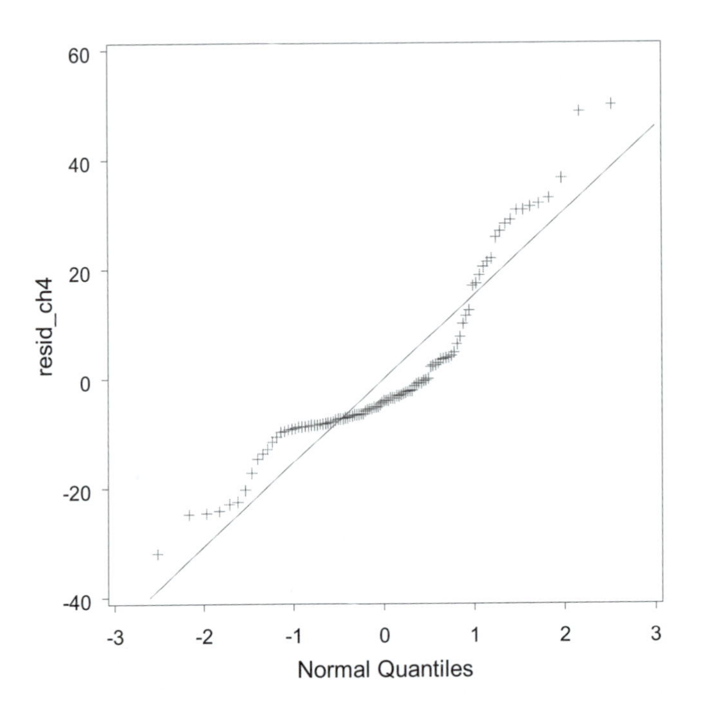

Figure 4.3: Q-Q plot of residuals.

4.7.5 Fitting regression models separately for each value of another variable

One common task is to perform identical analyses in several groups. Here, as an example, we consider separate linear regressions for each substance abuse group. This is a stratification approach to the interaction between `substance` and `age`. We show only the parameter estimates, using `ODS` to save and print them.

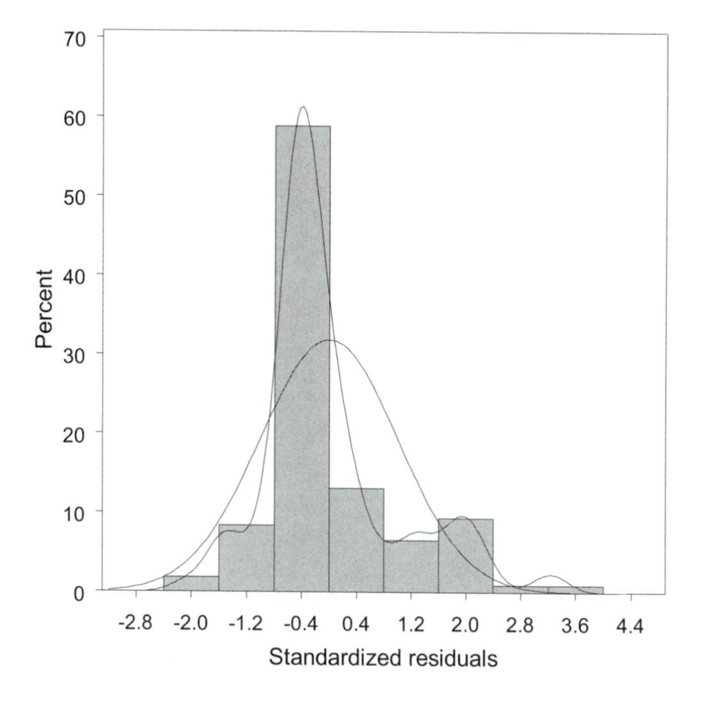

Figure 4.4: Empirical density of residuals, with superimposed normal density.

```
ods select none;
proc sort data=help;
  by substance;
run;

ods output parameterestimates=helpsubstparams;
proc glm data=help;
  by substance;
  model i1 = age / solution;
run;
ods select all;
```

```
options ls=64;
proc print data=helpsubstparams;
run;

  S       D  P
  U       e  a                    E
  B       p  r                    s
  S       e  a                    t           S        t
  T       n  m                    i           t        V      P
  A       d  e                    m           d        a      r
O N       e  t                    a           E        l      o
b C       n  e                    t           r        u      b
s E       t  r                    e           r        e      t

1 alcohol I1 Intercept  57.10998953  18.00474934    3.17 0.0032
2 alcohol I1 AGE        -0.71982952   0.45069028   -1.60 0.1195
3 cocaine I1 Intercept   5.25687957  11.52989056    0.46 0.6510
4 cocaine I1 AGE         0.11579282   0.32582541    0.36 0.7242
5 heroin  I1 Intercept  -7.77045212   8.59729637   -0.90 0.3738
6 heroin  I1 AGE         0.39337843   0.24179872    1.63 0.1150
```

The output shows that the `by` variable is included in the output dataset. So the estimated intercept for alcohol users is 57.1 drinks, declining by 0.7 drinks per year of age, just as we found in the interaction model. This stratified approach cannot show us that the alcohol and heroin groups are almost statistically significantly different from each other.

4.7.6 Two-way ANOVA

Is there a statistically significant association between gender and substance abuse group with depressive symptoms? We can make an interaction plot (6.1.11) by hand, as seen in Figure 4.5, or `proc glm` will make one automatically if the `ods graphics on` statement is issued. Note that we first reload the dataset so that the dataset again contains both genders.

```
libname k 'c:/book';

proc sort data=k.help;
   by substance female;
run;

ods select none;
proc means data=k.help;
  by substance female;
  var cesd;
  output out=helpmean mean=;
run;
ods select all;
```

```
axis1 minor=none;
symbol1 i=j v=none l=1 c=black w=5;
symbol2 i=j v=none l=2 c=black w=5;
proc gplot data=helpmean;
  plot cesd*substance = female / haxis=axis1 vaxis=axis1;
run; quit;
```

There are indications of large effects of gender and substance group, but little suggestion of interaction between the two. The same conclusion is reached in Figure 4.6, which displays boxplots by substance group and gender. The boxplot code is below.

```
data h2; set k.help;
  if female eq 1 then sex='F';
  else sex='M';
run;

proc sort data=h2; by sex; run;

symbol1 v = 'x' c = black;
proc boxplot data=h2;
  plot cesd * substance(sex) / notches boxwidthscale=1;
run;
```

The width of each box is proportional to the size of the sample, with the notches denoting confidence intervals for the medians, and X's marking the observed means.

Next, we proceed to formally test whether there is a significant interaction through a two-way analysis of variance (4.1.8). The Type III sums of squares table can be used to assess the interaction; we restrict output to this table to save space.

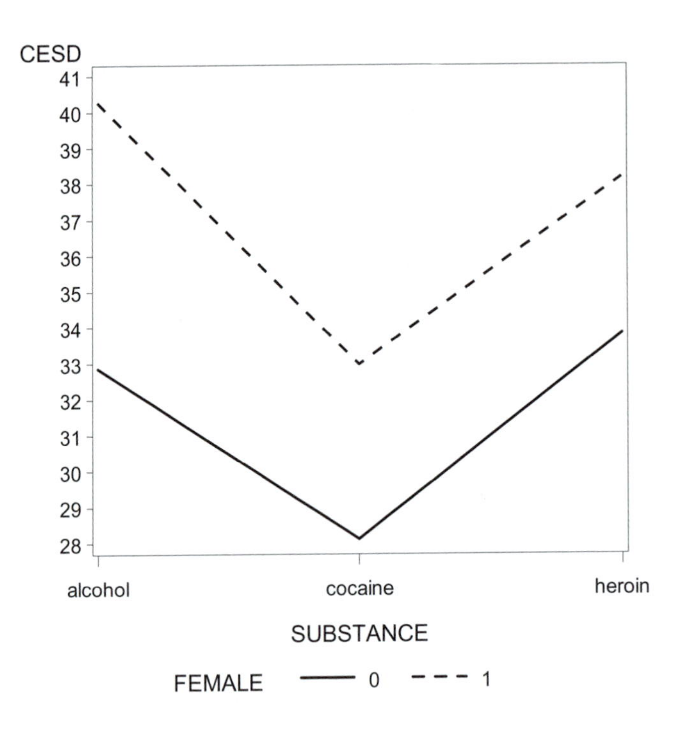

Figure 4.5: Interaction plot of CESD as a function of substance group and gender.

```
options ls=64;
ods select modelanova;
proc glm data=k.help;
  class female substance;
  model cesd = female substance female*substance / ss3;
run;

The GLM Procedure

Dependent Variable: CESD

Source                    DF     Type III SS    Mean Square

FEMALE                     1     2463.232928    2463.232928
SUBSTANCE                  2     2540.208432    1270.104216
FEMALE*SUBSTANCE           2      145.924987      72.962494

Source               F Value    Pr > F

FEMALE                 16.84    <.0001
SUBSTANCE               8.69    0.0002
FEMALE*SUBSTANCE        0.50    0.6075
```

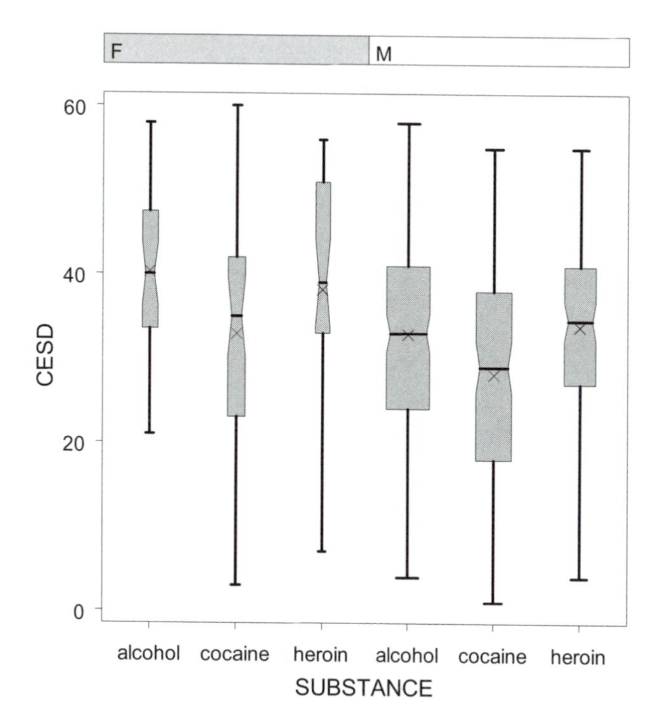

Figure 4.6: Boxplot of CESD as a function of substance group and gender.

There is little evidence (p=0.61) of an interaction, so this term can be dropped. We will estimate the reduced model.

```
options ls=64; /* stay in gray box */
ods select overallanova parameterestimates;
proc glm data=k.help;
  class female substance;
  model cesd = female substance / ss3 solution;
run;
```

```
The GLM Procedure

Dependent Variable: CESD

                                        Sum of
Source                      DF         Squares     Mean Square

Model                        3      5273.13263      1757.71088

Error                      449     65515.35744       145.91394

Corrected Total            452     70788.49007

Source                 F Value    Pr > F

Model                    12.05    <.0001

Error
Corrected Total
```

The model is very useful in explaining the variability in the data.

```
The GLM Procedure

Dependent Variable: CESD

                                                Standard
Parameter                    Estimate              Error    t Value

Intercept                 39.13070331 B       1.48571047      26.34
FEMALE      0             -5.61922564 B       1.33918653      -4.20
FEMALE      1              0.00000000 B        .               .
SUBSTANCE alcohol         -0.28148966 B       1.41554315      -0.20
SUBSTANCE cocaine         -5.60613722 B       1.46221461      -3.83
SUBSTANCE heroin           0.00000000 B        .               .

Parameter               Pr > |t|

Intercept                 <.0001
FEMALE      0             <.0001
FEMALE      1              .
SUBSTANCE alcohol         0.8425
SUBSTANCE cocaine         0.0001
SUBSTANCE heroin           .
```

Women exhibit more depressive symptoms (the effect when `female` is 0 is negative) while cocaine users are less depressed than heroin users. Heroin and alcohol users have similar depression levels.

The AIC criteria (4.2.3) can also be used to compare models. It is available in `proc reg` and `proc mixed` as well as some other procedures. Here we use `proc mixed`, omitting other output.

```
ods select fitstatistics;
proc mixed data=k.help method=ml;
  class female substance;
  model cesd = female|substance;
run; quit;

The Mixed Procedure

           Fit Statistics

-2 Log Likelihood              3537.8
AIC (smaller is better)        3551.8
AICC (smaller is better)       3552.1
BIC (smaller is better)        3580.6
```

```
ods select fitstatistics;
proc mixed data=k.help method=ml;
  class female substance;
  model cesd = female substance;
run; quit;
ods select all;

The Mixed Procedure

           Fit Statistics

-2 Log Likelihood              3538.8
AIC (smaller is better)        3548.8
AICC (smaller is better)       3549.0
BIC (smaller is better)        3569.4
```

The AIC criterion also suggests that the model without the interaction is most appropriate.

4.7.7 Multiple comparisons

We can also carry out multiple comparison (4.3.4) procedures to test each of the pairwise differences between substance abuse groups. This utilizes the `lsmeans` statement within `proc glm`.

```
ods select diff lsmeandiffcl lsmlines;
proc glm data=k.help;
  class substance;
  model cesd = substance;
   lsmeans substance / pdiff adjust=tukey cl lines;
run; quit;
ods select all;
```

```
The GLM Procedure
Least Squares Means
Adjustment for Multiple Comparisons: Tukey-Kramer

i/j              1             2             3

   1                        0.0009        0.9362
   2          0.0009                      0.0008
   3          0.9362        0.0008
```

This output shows the p-values for the pairwise comparisons, after adjustment.

```
The GLM Procedure
Least Squares Means
Adjustment for Multiple Comparisons: Tukey-Kramer

      Least Squares Means for Effect SUBSTANCE

            Difference           Simultaneous 95%
               Between        Confidence Limits for
i    j           Means         LSMean(i)-LSMean(j)

1    2        4.951829        1.753296     8.150361
1    3       -0.498086       -3.885335     2.889162
2    3       -5.449915       -8.950037    -1.949793
```

These are the adjusted confidence limits requested with the `cl` option.

```
The GLM Procedure
Least Squares Means
Adjustment for Multiple Comparisons: Tukey-Kramer

               Tukey-Kramer Comparison Lines for
                Least Squares Means of SUBSTANCE

LS-means with the same letter are not significantly different.

         CESD                  LSMEAN
       LSMEAN    SUBSTANCE     Number

A    34.87097    heroin             3
A
A    34.37288    alcohol            1

B    29.42105    cocaine            2
```

The above output demonstrates the results of the `lines` option using the `lsmeans` statement. The letter `A` shown on the left connecting the `heroin` and `alcohol` substances reflects the fact that there is not a statistically significant difference between these two groups. Since the `cocaine` substance has the letter `B` and no other group has one, the cocaine group is significantly different from each of the other groups. If instead the `cocaine` and `alcohol` substances *both* had a letter `B` attached, while the `heroin` and `alcohol` substances retained the letter `A` they have in the actual output, only the heroin and cocaine groups would differ significantly, while the alcohol group would differ from neither. This presentation becomes particularly useful as the number of groups increases.

All three pieces of output show that the alcohol group and heroin group both have significantly higher CESD scores than the cocaine group, but the alcohol and heroin groups do not significantly differ from each other. The 95% CI on the final comparison ranges from -2.9 to 3.9.

4.7.8 Contrasts

We can also fit contrasts (4.3.3) to test hypotheses involving multiple parameters. In this case, we can compare the CESD scores for the alcohol and heroin groups to the cocaine group. To allow checking the contrast, we use the `e` option to the `estimate` statement.

```
ods select contrastcoef estimates;
proc glm data=k.help;
  class female substance;
  model cesd = female substance;
  output out=outanova residual=resid_ch4anova;
  estimate 'A+H/2 = C?' substance 1 -2 1 / e divisor = 2;
run; quit;
ods select all;
```

In the code above, the `divisor` option allows you to use integer values to specify the contrast, but then return to the natural scale. This is especially useful if the contrast cannot be expressed as a decimal.

```
The GLM Procedure

Coefficients for Estimate A+H/2 = C?

                                Row 1

Intercept                           0

FEMALE    0                         0
FEMALE    1                         0

SUBSTANCE alcohol                 0.5
SUBSTANCE cocaine                  -1
SUBSTANCE heroin                  0.5
```

```
The GLM Procedure

Dependent Variable: CESD

                                               Standard
Parameter                       Estimate          Error    t Value

A+H/2 = C?                    5.46539238     1.21004493       4.52

Parameter                   Pr > |t|

A+H/2 = C?                    <.0001
```

There is a statistically significant difference in this one degree of freedom comparison ($p<0.0001$). The average in the alcohol and heroin groups is 5.5 units higher than in the cocaine group; cocaine users are less depressed.

Chapter 5

Regression generalizations and multivariate statistics

This chapter discusses many commonly used statistical models beyond linear regression and ANOVA. Most SAS procedures mentioned in this chapter support the `class` statement for categorical covariates (4.1.3).

5.1 Generalized linear models

Table 5.1 displays the options for specifying link functions and family of distributions for generalized linear models [31]. Description of several specific generalized linear regression models (e.g., logistic and Poisson) can be found in subsequent sections of this chapter. Some generalized linear models can also be accessed through one of the point-and-click interfaces (1.9).

```
proc genmod data=ds;
  model y = x1 ... xk / dist=familyname link=linkname;
run;
```

The `class` statement in `proc genmod` is more flexible than that available in many other procedures, notably `proc glm`. However, the default behavior is the same as for `proc glm` (see Section 4.1.3).

Table 5.1: Generalized Linear Model Distributions and Links in `proc genmod`

Distribution	Syntax
Gaussian	`dist=normal`
binomial	`dist=binomial`
gamma	`dist=gamma`
Poisson	`dist=poisson`
inverse Gaussian	`dist=igaussian`
Multinomial	`dist=multinomial`
Negative Binomial	`dist=negbin`
overdispersed	`dist=binomial` or `dist=multinomial` with `scale=deviance aggregate`, or `dist=poisson scale=deviance`

Note: The following links are available for all distributions: `identity`, `log`, or `power(`λ`)` (where λ is specified by the user). For dichotomous outcomes, complementary log-log (`link=cloglog`), logit (`link=logit`), or probit (`link=probit`) are additionally available. For multinomial distributed outcomes, cumulative complementary log-log (`link=cumcll`), cumulative logit (`link=cumlogit`), or cumulative probit (`link=cumprobit`) are available. Once the family and link functions have been specified, the variance function is implied (with the exception of the `quasi` family). Overdispersion is implemented using the `scale` option to the model statement. To allow overdispersion in Poisson, binomial, or multinomial models, use the option `scale=deviance`; the additional `aggregate` option is required for the binomial and multinomial. Any valid link listed above may be used.

5.1.1 Logistic regression model

Example: See 5.6.1

See also 5.3.2 (exact logistic regression) and 5.3.3 (conditional logistic regression)

```
proc logistic data=ds;
  model y = x1 ... xk / or cl;
run;
```

or

```
proc logistic data=ds;
  model y(event='1') = x1 ... xk;
run;
```

or

```
proc logistic data=ds;
  model r/n = x1 ... xk / or cl; /* events/trials syntax */
run;
```

or

```
proc genmod data=ds;
  model y = x1 ... xk / dist=binomial link=logit;
run;
```

While both procedures will fit logistic regression models, `proc logistic` is likely to be more useful for ordinary logistic regression than `proc genmod`. The former allows options such as those printed above in the first `model` statement, which produce the odds ratios (and their confidence limits) associated with the log-odds estimated by the model. It also produces the area under the ROC curve (the so-called "c" statistic) by default (see also 6.1.17). Both procedures allow the logit, probit, and complementary log-log links, through the `link` option to the `model` statement; `proc genmod` must be used if other link functions are desired.

The `events/trials` syntax can be used to save storage space for data. In this case, observations with the same covariate values are stored as a single line of data, with the number of observations recorded in one variable (`trials`) and the number with the outcome in another (`events`).

The output from `proc logistic` and `proc genmod` prominently display the level of `y` that is being predicted. The `descending` option to the `proc` statement will reverse the order. Alternatively, the `model` statement in `proc logistic` allows you to specify the target level as shown in the second set of code.

The `class` statement in `proc genmod` is more flexible than that available in many other procedures, notably `proc glm`. Importantly, the default behavior is different than in `proc glm` (see Section 4.1.3).

5.1.2 Poisson model

See also 5.1.3 (zero-inflated Poisson) *Example:* See 5.6.2

```
proc genmod data=ds;
  model y = x1 ... xk / dist=poisson;
run;
```

The default output from `proc genmod` includes useful methods to assess fit.

5.1.3 Zero-inflated Poisson model

Example: See 5.6.3

Zero-inflated Poisson models can be used for count outcomes that generally follow a Poisson distribution but for which there are (many) more observed counts of 0 than would be expected. These data can be seen as deriving from a mixture distribution of a Poisson and a degenerate distribution with point mass at zero (see also 5.1.5, zero-inflated negative binomial).

```
proc genmod data=ds;
  model y = x1 ... xk / dist=zip;
  zeromodel x2 ... xp;
run;
```

The Poisson rate parameter of the model is specified in the `model` statement, with a default log link and alternate link functions available as described in Table 5.1. The extra zero-probability is modeled as a logistic regression of the covariates specified in the `zeromodel` statement. Support for zero-inflated Poisson models is also available within `proc countreg`.

5.1.4 Negative binomial model

See also 5.1.5 (zero-inflated negative binomial) *Example:* See 5.6.4

```
proc genmod data=ds;
  model y = x1 ... xk / dist=negbin;
run;
```

5.1.5 Zero-inflated negative binomial model

Zero-inflated negative binomial models can be used for count outcomes that generally follow a negative binomial distribution but for which there are (many) more observed counts of 0 than would be expected. These data can be seen as deriving from a mixture distribution of a negative binomial and a degenerate distribution with point mass at zero (see also zero-inflated Poisson, 5.1.3).

```
proc countreg data=help2;
  model y = x1 ... xk / dist=zinb;
  zeromodel y ~ x2 ... xp;
run;
```

The negative binomial rate parameter of the model is specified in the `model` statement. The extra zero-probability is modeled as a function of the covariates specified after the ~ in the `zeromodel` statement.

5.1.6 Ordered multinomial model

Example: See 5.6.6

```
proc genmod data=ds;
  model y = x1 ... xk / dist=multinomial;
run;
```

or

```
proc logistic data=ds;
  model y = x1 ... xk / link=cumlogit;
run;
```

The genmod procedure utilizes a cumulative logit link by default when the dist is multinomial, comparing each level of the outcome with all lower levels. The model implies the proportional odds assumption. The cumulative probit model is available with the link=cprobit option to the model statement in proc genmod. The proc logistic implementation provides a score test for the proportional odds assumption.

5.1.7 Generalized (nominal outcome) multinomial logit

Example: See 5.6.7

```
proc logistic data=ds;
  model y = x1 ... xk / link=glogit;
run;
```

Each level is compared to a reference level, which can be chosen using the ref option, e.g., model y(ref='0') = x1 / link=glogit.

5.2 Models for correlated data

There is extensive support within SAS for correlated data regression models, including repeated measures, longitudinal, time series, clustered, and other related methods. Throughout this section we assume that repeated measurements are taken on a subject or cluster denoted by variable id.

5.2.1 Linear models with correlated outcomes

Example: See 5.6.10

```
proc mixed data=ds;
  class id;
  model y = x1 ... xk;
  repeated / type=vartype subject=id;
run;
```

or

```
proc mixed data=ds;
  class id;
  model y = x1 ... xk / outpm=dsname;
  repeated ordervar / type=covtypename subject=id;
run;
```

The solution option to the model statement can be used to get fixed effects parameter estimates in addition to ANOVA tables. The repeated ordervar syntax is used when observations within a cluster are (a) ordered (as in repeated measurements), (b) the placement in the order affects the covariance structure (as in most structures other than independence and compound symmetry), and (c) observations may be missing from the beginning or middle of the order. Predicted values for observations can be found using the outpm option to the model statement as demonstrated in the second block of code. To add to the outpm dataset the outcomes and transformed residuals (scaled by the inverse Cholesky root of the marginal covariance matrix), add the vciry option to the model statement.

The structure of the covariance matrix of the observations is controlled by the type option to the repeated statement. As of SAS 9.2, there are 36 available structures. Particularly useful options include un (unstructured), cs (compound symmetry), and ar(1) (first-order autoregressive). The full list is available through the online help: Contents; SAS Products; SAS Procedures; MIXED; Syntax; REPEATED.

5.2.2 Linear mixed models with random intercepts

See also 5.2.3 (random slope models), 5.2.4 (random coefficient models), and 7.2.2 (empirical power calculations)

```
proc mixed data=ds;
  class id;
  model y = x1 ... xk;
  random int / subject=id;
run;
```

The solution option to the model statement may be required to get fixed effects parameter estimates in addition to ANOVA tables. The random statement describes the design matrix for the random effects. Unlike the fixed effects design matrix, specified as usual with the model statement, the random effects design matrix includes a random intercept only if it is specified as above. The predicted random intercepts can be printed with the solution option to the random statement and saved into a dataset using the ODS system, e.g., an ods output solutionr=reffs statement. Predicted values for observations can be found using the outp=datasetname and outpm=datasetname options to the model statement; the outp dataset includes the predicted random effects in the predicted values while the outpm predictions include only the fixed effects.

5.2.3 Linear mixed models with random slopes

Example: See 5.6.11

See also 5.2.2 (random intercept models) and 5.2.4 (random coefficient models)

```
proc mixed data=ds;
  class id;
  model y = time x1 ... xk;
  random int time / subject=id type=covtypename;
run;
```

The `solution` option to the `model` statement can be used to get fixed effects parameter estimates in addition to ANOVA tables. Random effects may be correlated with each other (though not with the residual errors for each observation). The structure of the the covariance matrix of the random effects is controlled by the `type` option to the `random` statement. The option most likely to be useful is `type=un` (unstructured); by default, `proc mixed` uses the variance component (`type=vc`) structure, in which the random effects are uncorrelated with each other. The predicted random effects can be printed with the `solution` option to the `random` statement and saved into a dataset using the `ODS` system, e.g., an `ods output solutionr=reffs` statement. Predicted values for observations can be found using the `outp=datasetname` and `outpm=datasetname` options to the `model` statement; the `outp` dataset includes the predicted random effects in the predicted values while the `outpm` predictions include only the fixed effects.

5.2.4 More complex random coefficient models

We can extend the random effects models introduced in 5.2.2 and 5.2.3 to 3 or more subject-specific random parameters (e.g., a quadratic growth curve or spline/"broken stick" model [14]). In the below, we use `time1` and `time2` to refer to two generic functions of time.

```
proc mixed data=ds;
  class id;
  model y = time1 time2 x1 ... xk;
  random int time1 time2 / subject=id type=covtypename;
run;
```

The `solution` option to the `model` statement can be used to get fixed effects parameter estimates in addition to ANOVA tables. Random effects may be correlated with each other, though not with the residual errors for each observation. The structure of the covariance matrix of the random effects is controlled by the `type` option to the `random` statement. The option most likely to be useful is `type=un` (unstructured); by default, `proc mixed` uses the variance component (`type=vc`) structure, in which the random effects are uncorrelated with each other. The predicted random effects can be printed with the `solution` option to the `random` statement and saved into a dataset using the `ODS` system, e.g., `ods output solutionr=reffs`. Predicted values for observations can be found using the `outp` and `outpm` options to the `model` statement; the `outp` dataset includes the predicted random effects in the predicted values while the `outpm` predictions include only the fixed effects.

5.2.5 Multilevel models

Studies with multiple levels of clustering can be fit in SAS. In a typical example, a study might include schools (as one level of clustering) and classes within schools (a second level of clustering), with individual students within the classrooms providing a response. Generically, we refer to $level_l$ variables which are identifiers of cluster membership at level l. Random effects at different levels are assumed to be uncorrelated with each other.

```
proc mixed data=ds;
  class id;
  model y = x1 ... xk;
  random int / subject=level1;
  random int / subject=level2;
run;
```

Each `random` statement uses a `subject` option to describe a different clustering structure in the data. There is no theoretical limit to the complexity of the structure or the number of `random` statements, but practical difficulties in fitting the models may be encountered.

5.2.6 Generalized linear mixed models

Example: See 5.6.13 and 7.2.2

```
proc glimmix data=ds;
  model y = time x1 ... xk / dist=familyname link=linkname;
  random int time / subject=id type=covtypename;
run;
```

Observations sharing a value for `id` are correlated; otherwise, they are assumed independent. Random effects may be correlated with each other (though not with the residual errors for each observation). The structure of the the covariance matrix of the random effects is controlled by the `type` option to the `random` statement. There are many available structures. The one most likely to be useful is `un` (unstructured). The full list is available through the online help: Contents; SAS Products; SAS Procedures; GLIMMIX; Syntax; RANDOM. As of SAS 9.2, all of the distributions and links shown in Table 5.1 are available, and additionally `dist` can be `beta`, `exponential`, `geometric`, `lognormal`, or `tcentral` (t distribution). An additional link is available for nominal categorical outcomes: `glogit` the generalized logit. Note that the default fitting method relies on an approximation to an integral. The `method=laplace` option to the `proc glimmix` statement will use a numeric integration (this is likely to be time-consuming).

For SAS 9.1 users, `proc glimmix` is available from SAS Institute as a free add-on package: `http://support.sas.com/rnd/app/da/glimmix.html`.

5.2.7 Generalized estimating equations

Example: See 5.6.12

```
proc genmod data=ds;
  model y = x1 ... xk;
  repeated / subject=id type=corrtypename;
run;
```

The `repeated ordervar` syntax should be used when observations within a cluster are (a) ordered (as in repeated measurements), (b) the placement in the order affects the covariance structure (as in most structures other than independence and compound symmetry), and (c) observations may be missing from the beginning or middle of the order. The structure of the working covariance matrix of the observations is controlled by the `type` option to the `repeated` statement. The available correllation types as of SAS 9.2 include `ar` (first-order autoregressive), `exch` (exchangeable), `ind` (independent), `mdep(m)` (m-dependent), `un` (unstructured), and `user` (a fixed, user-defined correlation matrix).

5.2.8 Time-series model

Time-series modeling is an extensive area with a specialized language and notation. We demonstrate fitting a simple ARIMA (autoregressive integrated moving average) model for the first difference, with first-order auto-regression and moving averages.

The procedures to fit time series data are included in the SAS/ETS package. These provide extensive support for time series analysis. However, it is also possible to fit simple auto-regressive models using `proc mixed`. We demonstrate the basic use of `proc arima` (from SAS/ETS).

```
proc arima data=ds;
  identify var=x(1);
  estimate p=1 q=1;
run;
```

In proc arima, the variable to be analyzed is specified in the identify statement, with differencing specified in parentheses. The estimate statement specifies the order of the auto-regression (p) and moving average (q). Prediction can be accomplished via the forecast statement.

5.3 Further generalizations to regression models

5.3.1 Proportional hazards (Cox) regression model

Example: See 5.6.14

Survival or failure time data, typically consist of the time until the event, as well as an indicator of whether the event was observed or censored at that time. Throughout, we denote the time of measurement with the variable time and censoring with a dichotomous variable cens = 1 if censored, or = 0 if observed. Other entries related to survival analysis include 3.4.4 (logrank test) and 6.1.18 (Kaplan–Meier plot).

```
proc phreg data=ds;
  model time*cens(1) = x1 ... xk;
run;
```

SAS supports time varying covariates using programming statements within proc phreg. The class statement in proc genmod is more flexible than that available in many other procedures, notably proc glm. However, the default behavior is the same as for proc glm (see Section 4.1.3).

5.3.2 Exact logistic regression

See also 5.1.1 (logistic regression) and 5.3.3 (conditional logistic regression)

```
proc logistic data=ds;
  model y = x1 ... xk;
  exact intercept x1;
run;
```

An exact test is generated for each variable listed in the exact statement, including if desired the intercept, as shown above. Not all covariates in the model statement need be included in the exact statement, but all covariates in the exact statement must be included in the model statement.

5.3.3 Conditional logistic regression model

See also 5.1.1 (logistic regression) and 5.3.2 (exact logistic regression)

```
proc logistic data=ds;
  strata id;
  model y = x1 ... xk;
run;
```

or

```
proc logistic data=ds;
  strata id;
  model y = x1 ... xk;
  exact intercept x1;
run;
```

The variable `id` identifies strata or matched sets of observations. An exact model can be fit using the `exact` statement with list of covariates to be assessed using an exact test, including the intercept, as shown above.

5.3.4 Log-linear model

Log-linear models are a flexible approach to analysis of categorical data [2]. A loglinear model of a three-dimensional contingency table denoted by X_1, X_2, and X_3 might assert that the expected counts depend on a two-way interaction between the first two variables, but that X_3 is independent of all the others:

$$log(m_{ijk}) = \mu + \lambda_i^{X_1} + \lambda_j^{X_2} + \lambda_{ij}^{X_1,X_2} + \lambda_k^{X_3}$$

```
proc catmod data=ds;
  weight count;
  model x1*x2*x3 =_response_ / pred;
  loglin x1|x2 x3;
run;
```

The variables listed in the `model` statement above describe the n-way table to be analyzed; the term `_response_` is a required keyword indicating a log-linear model. The `loglin` statement specifies the dependence assumptions. The `weight` statement is optional. If used, the `count` variable should contain the cell counts and can be used if the analysis is based on a summary dataset.

5.3.5 Nonlinear least squares model

Nonlinear least squares models [45] can be fit flexibly within SAS. As an example, consider the income inequality model described by Sarabia and colleagues [41]:

$$Y = (1 - (1 - X)^p)^{(1/p)}$$

We provide a starting value (0.5) within the interior of the parameter space.

```
proc nlin data=ds;
  parms p=0.5;
  model y = (1 - ((1-x)**p))**(1/p);
run;
```

5.3.6 Generalized additive model

Example: See 5.6.8

```
proc gam data=ds;
  model y = spline(x1, df) loess(x2) spline2(x3, x4) ...
    param(x5 ... xk);
run;
```

Specification of a spline or lowess term for variable `x1` is given by `spline(x1)` or `loess(x1)`, respectively, while a bivariate spline fit can be included using `spline2(x1, x2)`. The degrees of freedom can be specified as in `spline(x1, df)`, following a comma in the variable function description, or estimated from the model using generalized cross-validation by including the `method=gcv` option in the model statement. If neither is specified, the default degrees of freedom of 4 is used. Any variables included in `param()` are fit as linear predictors with the usual syntax (4.1.5).

5.3.7 Model selection

There are many methods for automating the selection of covariates to include in models. The most common of these may be the stepwise methods, which build models by adding variables one at a time, removing them one at a time, or both adding and removing. Another well-known method is the lasso, due to the work by Tibshirani [49]. In most applications, we recommend approaching model selection with a more intelligent approach, but when confronted with large potential predictor pools, model selection methods can be useful.

```
proc reg data=ds;
  model y = x1 ... xk / selection=stepwise;
run;
```

or

```
proc glmselect data=ds;
  class x2 xk;
  model y = x1 ... xk / selection=stepwise;
run;
```

or

```
proc logistic data=ds;
  class x2 xk;
  model y = x1 ... xk / selection=stepwise;
run;
```

The `reg` procedure will perform `stepwise`, `forward`, and `backward` selection methods. It will also perform a variety of other methods based on maximizing various model-fit statistics. The `glmselect` procedure, available starting with SAS 9.2, is for general linear models, meaning for models with uncorrelated normal errors. Unlike `proc reg`, it will treat categorical values appropriately, meaning that all of the indicator variables (implied by the class statement, 4.7.6) will be included or excluded at each step. Available methods are `stepwise`, `lasso`, `lars`, `forward`, and `backward`. For logistic regression and related models, `proc logistic` can perform the `stepwise`, `forward`, `backward`, and `score` methods. Categorical variables listed in the `class` statement are handled appropriately, though the `score` method cannot be used with them. The `score` method fits all possible models with each given number of predictors. This can become prohibitively costly in compute time, and can be controlled through the `best`, `start`, and `stop` options. For each of `proc reg`, `proc glmselect`, and `proc logistic`, the entry probability (forward and stepwise methods) is set with the `slentry` option and removal probability (backward and stepwise methods) is set with the `slstay` option.

5.3.8 Quantile regression model

Example: See 5.6.5

Quantile regression predicts changes in the specified quantile of the outcome variable per unit change in the predictor variables; analogous to the change in the mean predicted in least squares regression. If the quantile so predicted is the median, this is equivalent to minimum absolute deviation regression (as compared to least squares regression minimizing the squared deviations).

```
proc quantreg data=ds;
  model y = x1 ... xk / quantile=0.75;
run;
```

The `quantile` option specifies which quantile is to be estimated (here the 75th percentile). Median regression (i.e., `quantile=0.50`) is performed by default. If multiple quantiles are included (separated by commas) then they are estimated simultaneously, but standard errors and tests are only carried out when a single quantile is provided.

5.3.9 Ridge regression model

```
proc reg data=ds ridge=a to b by c;
  model y = x1 ... xk;
run;
```

Each of the values `a, a+c, a + 2c`, ..., `b` is added to the diagonal of the cross-product matrix of $X_1, \ldots, X_k$. Ridge regression estimates are the least squares estimates obtained using this new cross-product matrix.

5.3.10 Bayesian regression methods

Example: See 5.6.15

Bayesian methods are increasingly commonly utilized. An overview of Bayesian methods in SAS is discussed in the online help: Contents; SAS Products; SAS/STAT; SAS/STAT User's Guide, under "Introduction to Bayesian Analysis Procedures." Here we show how to use Bayesian generalized linear models through Markov Chain Monte Carlo (MCMC) in `proc genmod` and how to fit a simple linear regression using MCMC in the far more general `proc mcmc`.

```
proc genmod data=ds;
  model y = x1 ... xk / dist=familyname link=linkname;
  bayes;
run;
```

or

```
proc mcmc data=ds;
  parms beta0 0 beta1 0 sigmasq 1;
  prior beta0 beta1 ~ normal(mean=0, var=100000);
  prior sigmasq ~ igamma(shape=0.1, scale=10);
  mu = beta0 + beta1*x;
  model y ~ normal(mean=mu, var=sigmasq);
run;
```

The `lifereg` and `phreg` procedures will perform Bayesian analyses of survival data. The `mcmc` procedure is general enough that it can fit generalized linear mixed models such as those described in 5.2.6 and nonlinear models such as 5.3.5 as well as many other models not available in specific procedures. A host of options are available for diagnostic about the MCMC iterations, as is necessary for responsible data analysis.

5.3.11 Complex survey design

The appropriate analysis of sample surveys requires incorporation of complex design features, including stratification, clustering, weights, and finite population correction. These can be addressed in SAS for many common models. In this example, we assume that there are variables `psuvar` (cluster or PSU), `stratum` (stratification variable), and `wt` (sampling weight). Code examples are given to estimate the mean of a variable `x1` as well as a linear regression model.

```
proc surveymeans data=ds rate=fpcvar;
  cluster psuvar;
  strata stratum;
  weight wt;
  var x1 ... xk;
run;
```

or

```
proc surveyreg data=ds rate=fpcvar;
  cluster psuvar;
  strata stratum;
  weight wt;
  model y = x1 ... xk;
run;
```

The `surveymeans` and `surveyreg` procedures account for complex survey designs with equivalent functionality to `means` and `reg`, respectively. Other survey procedures in SAS include `surveyfreq` and `surveylogistic`, which emulate procedures `freq` and `logistic`. The survey procedures share a `strata` statement to describe the stratification variables, a `weight` statement to describe the sampling weights, a `cluster` statement to specify the primary sampling unit (PSU) or cluster, and a `rate` option (for the `proc` statement) to specify a finite population correction as a count or dataset. Additional options allow specification of the total number of PSUs or a dataset with the number of PSUs in each stratum.

5.4 Multivariate statistics and discriminant procedures

This section includes a sampling of commonly used multivariate, clustering methods, and discriminant procedures [30, 47].

Summaries of these topics and how to implement related methods are discussed in the online help: Contents; SAS Products; SAS/STAT; SAS/STAT User's Guide under the headings "Introduction to Multivariate Procedures," "Introduction to Clustering Procedures," and "Introduction to Discriminant Procedures."

5.4.1 Cronbach's α

Example: See 5.6.16

Cronbach's α is a statistic that summarizes the internal consistency and reliability of a set of items comprising a measure.

```
proc corr data=ds alpha nomiss;
  var x1 ... xk;
run;
```

The `nomiss` option is required so that only observations with all variables observed are included.

5.4.2 Factor analysis

Example: See 5.6.17

Factor analysis is used to explain variability of a set of measures in terms of underlying unobservable factors. The observed measures can be expressed as linear combinations of the factors, plus random error. Factor analysis is often used as a way to guide the creation of summary scores from individual items.

```
proc factor data=ds nfactors=k rotate=rotatemethod;
  var x1 ... xk;
run;
```

The `nfactors` option forces k factors; the default is the number of variables. There are 25 rotation methods available. The method for fitting can be controlled through the `method` option to `proc factor`.

5.4.3 Linear discriminant analysis

Example: See 5.6.18

Linear (or Fisher) discriminant analysis is used to find linear combinations of variables that can separate classes.

```
proc discrim data=ds;
  class y;
  var x1 ... xk;
run;
```

The classification results and posterior probabilities can be output to a dataset using the `out` option to `proc discrim`.

5.4.4 Hierarchical clustering

Example: See 5.6.19

Many techniques exist for grouping similar variables or similar observations. An overview of clustering methods is discussed in the online help: Contents; SAS Products; SAS/STAT; SAS/STAT User's Guide, under the heading "Introduction to Clustering Procedures." We show one procedure to cluster variables and one to cluster observations.

```
proc varclus data=ds outtree=treedisp;
  var x1 ... xk;
run;
```

or

```
proc cluster data=ds method=methodname outtree=treedisp;
  var x1 ... xk;
run;
```

The `varclus` procedure is one of the procedures that can be used to cluster variables. The `cluster` procedure is one of the procedures that can cluster observations. For the `cluster` procedure, the `method` option is required. There are 11 methods available.

The `tree` procedure can be used to plot tree diagrams from hierarchical clustering results. The clustering procedures have an `outtree` option to write output that the `tree` procedure accepts as input.

5.5 Further resources

Many of the topics covered in this chapter are active areas of statistical research and many foundational articles are still useful. Here we provide references to texts which serve as accessible references.

Dobson and Barnett [10] is an accessible introduction to generalized linear models, while McCullagh and Neldor's [31] work remains a classic. Agresti [2] describes the analysis of categorical data.

Fitzmaurice, Laird, and Ware [14] provide an accessible overview of mixed effects methods while West, Welch, and Galecki [56] review these methods for a variety of statistical packages. The text by Hardin and Hilbe [18] provides a review of generalized estimating equations.

Collett [6] presents an accessible introduction to survival analysis.

Särndal, Swensson, and Wretman [42] provide a readable overview of the analysis of data from complex surveys.

Manly [30] and Tabachnick and Fidell [47] provide a comprehensive introduction to multivariate statistics.

5.6 HELP examples

To help illustrate the tools presented in this chapter, we apply many of the entries to the HELP data. SAS code can be downloaded from `http://www.math.smith.edu/sas/examples`.

```
libname k "c:/book";

data help;
  set k.help;
run;
```

In general, SAS output is lengthy. We annotate the full output with named ODS objects for logistic regression (Section 5.6.1), as discussed in Section 1.7. In some examples we provide the bulk of results, but we use ODS to reduce the output to a few key elements for the sake of brevity for most entries.

5.6.1 Logistic regression

In this example, we fit a logistic regression (5.1.1) where we model the probability of being homeless (spending one or more nights in a shelter or on the street in the past six months) as a function of predictors.

We can specify the `param` option to make SAS use the desired reference category (see 4.1.3).

```
options ls=64;  /* keep output in grey box */
proc logistic data=help descending;
  class substance (param=ref ref='alcohol');
  model homeless = female i1 substance sexrisk indtot;
run;
```

SAS produces a large number of distinct pieces of output by default. Here we reproduce the ODS name of each piece of output. These can be found by running `ods trace on / listing` before the procedure, as introduced in Section 1.7. Each ODS object can also be saved as a SAS dataset using these names with the `ods output` statement as shown later in this chapter.

First, SAS reports basic information about the model and the data in the ODS `modelinfo` output.

```
The LOGISTIC Procedure

                 Model Information

Data Set                             WORK.HELP
Response Variable                    HOMELESS
Number of Response Levels            2
Model                                binary logit
Optimization Technique               Fisher's scoring
```

Then SAS reports the number of observations read and used, in the ODS `nobs` output. Note that missing data will cause these numbers to differ. Subsetting with the `where` statement (1.6.3) will cause the number of observations displayed here to differ from the number in the dataset.

```
Number of Observations Read           453
Number of Observations Used           453
```

The ODS `responseprofile` output tabulates the number of observations with each outcome, and, importantly, reports which level is being modeled as the event.

```
             Response Profile

 Ordered                        Total
   Value     HOMELESS       Frequency

       1            1             209
       2            0             244

Probability modeled is HOMELESS=1.
```

The ODS `classlevelinfo` output shows the coding for each `class` variable. Here, `alcohol` is the reference category.

```
        Class Level Information

                              Design
Class             Value      Variables

SUBSTANCE         alcohol      0      0
                  cocaine      1      0
                  heroin       0      1
```

Whether the model converged is reported in the ODS `convergencestatus` output.

```
                Model Convergence Status

         Convergence criterion (GCONV=1E-8) satisfied.
```

Akaike Information Criterion (AIC) and other fit statistics are produced in the ODS `fitstatistics` output. We can infer that all are reported in the smaller-is-better format.

```
         Model Fit Statistics

                                 Intercept
                   Intercept           and
Criterion               Only    Covariates

AIC                  627.284       590.652
SC                   631.400       619.463
-2 Log L             625.284       576.652
```

Tests reported in the ODS `globaltests` output assess the joint null hypothesis that all parameters except the intercept equal 0.

```
        Testing Global Null Hypothesis: BETA=0

Test                 Chi-Square       DF     Pr > ChiSq

Likelihood Ratio        48.6324        6         <.0001
Score                   45.6522        6         <.0001
Wald                    40.7207        6         <.0001
```

The ODS `type3` output contains tests for adding each covariate (including joint tests for `class` variables with 2 or more values) to a model containing all other covariates. The two parameters needed for the `substance` variable donot jointly reach the 0.05 level of statistical significance.

```
          Type 3 Analysis of Effects

                              Wald
Effect           DF     Chi-Square    Pr > ChiSq

FEMALE            1         1.0831        0.2980
I1                1         7.6866        0.0056
SUBSTANCE         2         4.2560        0.1191
SEXRISK           1         3.4959        0.0615
INDTOT            1         8.2868        0.0040
```

The ODS `parameterestimates` output shows the maximum likelihood estimates of the parameters, their standard errors, and Wald statistics and tests for the null hypothesis that the parameter value is 0. Note that in this table, as opposed to the previous one, each level (other than the referent) of any `class` variable is reported separately. We see that the difference between cocaine and alcohol users approaches the 0.05 significance level, while heroin and cocaine users are more similar.

Analysis of Maximum Likelihood Estimates

Parameter	DF	Estimate	Standard Error	Wald Chi-Square	Pr > ChiSq
Intercept	1	-2.1319	0.6335	11.3262	0.0008
FEMALE	1	-0.2617	0.2515	1.0831	0.2980
I1	1	0.0175	0.00631	7.6866	0.0056
SUBSTANCE cocaine	1	-0.5033	0.2645	3.6206	0.0571
SUBSTANCE heroin	1	-0.4431	0.2703	2.6877	0.1011
SEXRISK	1	0.0725	0.0388	3.4959	0.0615
INDTOT	1	0.0467	0.0162	8.2868	0.0040

The ODS `oddsratios` output shows the exponentiated parameter estimates and associated confidence limits.

Odds Ratio Estimates

Effect	Point Estimate	95% Wald Confidence Limits	
FEMALE	0.770	0.470	1.260
I1	1.018	1.005	1.030
SUBSTANCE cocaine vs alcohol	0.605	0.360	1.015
SUBSTANCE heroin vs alcohol	0.642	0.378	1.091
SEXRISK	1.075	0.997	1.160
INDTOT	1.048	1.015	1.082

The ODS `association` output shows various other statistics. The area under the Receiver Operating Characteristic curve is denoted by "`c`."

Association of Predicted Probabilities and Observed Responses

Percent Concordant	67.8	Somers' D	0.360
Percent Discordant	31.8	Gamma	0.361
Percent Tied	0.4	Tau-a	0.179
Pairs	50996	c	0.680

If the parameter estimates are desired as a dataset, `ODS` can be used in SAS.

```
ods exclude all;
ods output parameterestimates=helplogisticbetas;
proc logistic data=help descending;
  class substance (param=ref ref='alcohol');
  model homeless = female i1 substance sexrisk indtot;
run;

ods exclude none;
options ls=64;
proc print data=helplogisticbetas;
run;
```

```
                  Class                                       Prob
Obs Variable      Val0     DF Estimate    StdErr  WaldChiSq  ChiSq

  1 Intercept              1  -2.1319    0.6335     11.3262 0.0008
  2 FEMALE                 1  -0.2617    0.2515      1.0831 0.2980
  3 I1                     1   0.0175   0.00631      7.6866 0.0056
  4 SUBSTANCE cocaine      1  -0.5033    0.2645      3.6206 0.0571
  5 SUBSTANCE heroin       1  -0.4431    0.2703      2.6877 0.1011
  6 SEXRISK                1   0.0725    0.0388      3.4959 0.0615
  7 INDTOT                 1   0.0467    0.0162      8.2868 0.0040
```

5.6.2 Poisson regression

In this example we fit a Poisson regression model (5.1.2) for `i1`, the average number of drinks per day in the 30 days prior to entering the detox center.

```
options ls=64;
ods exclude modelinfo nobs classlevels convergencestatus;
proc genmod data=help;
  class substance;
  model i1 = female substance age / dist=poisson;
run;
```

```
The GENMOD Procedure

              Criteria For Assessing Goodness Of Fit

Criterion                    DF           Value        Value/DF

Deviance                    448       6713.8986         14.9864
Scaled Deviance             448       6713.8986         14.9864
Pearson Chi-Square          448       7933.2027         17.7080
Scaled Pearson X2           448       7933.2027         17.7080
Log Likelihood                       16385.3197
Full Log Likelihood                  -4207.6544
AIC (smaller is better)               8425.3089
AICC (smaller is better)              8425.4431
BIC (smaller is better)               8445.8883
```

It is always important to check assumptions for models. This is particularly true for Poisson models, which are quite sensitive to model departures. The output includes several assessments of goodness of fit by default. The deviance value per degree of freedom (DF) is high (14.99)— models that fit well have deviance per DF near 1.

In the following output, the chi-square statistic and the confidence limits for the parameter estimates have been removed.

```
        Analysis Of Maximum Likelihood Parameter Estimates

                                                Standard
Parameter                 DF     Estimate          Error     Pr > ChiSq

Intercept                  1       1.7767         0.0582         <.0001
FEMALE                     1      -0.1761         0.0280         <.0001
SUBSTANCE   alcohol        1       1.1212         0.0339         <.0001
SUBSTANCE   cocaine        1       0.3040         0.0381         <.0001
SUBSTANCE   heroin         0       0.0000         0.0000              .
AGE                        1       0.0132         0.0015         <.0001
Scale                      0       1.0000         0.0000

NOTE: The scale parameter was held fixed.
```

The results show that alcohol users had $e^{1.1212} = 3.07$ times as many drinks as heroin users in the month before entering detox, holding age and gender constant.

5.6.3 Zero-inflated Poisson regression

A zero-inflated Poisson regression model (5.1.3) might fit better than the Poisson model shown in the previous section. Here we retain the model for the number of drinks, but allow for an extra probability of 0 drinks which might differ by gender.

```
options ls=64;
ods select parameterestimates zeroparameterestimates;
proc genmod data=help;
  class substance;
  model i1 = female substance age / dist=zip;
  zeromodel female;
run;
```

In the following output, the chi-square statistic and the confidence limits for the parameter estimates have been removed.

```
The GENMOD Procedure

      Analysis Of Maximum Likelihood Parameter Estimates

                                            Standard
Parameter                DF     Estimate       Error    Pr > ChiSq

Intercept                 1       2.2970      0.0599        <.0001
FEMALE                    1      -0.0680      0.0280        0.0153
SUBSTANCE    alcohol      1       0.7609      0.0336        <.0001
SUBSTANCE    cocaine      1       0.0362      0.0381        0.3427
SUBSTANCE    heroin       0       0.0000      0.0000         .
AGE                       1       0.0093      0.0015        <.0001
Scale                     0       1.0000      0.0000

NOTE: The scale parameter was held fixed.
```

```
        Analysis Of Maximum Likelihood Zero
           Inflation Parameter Estimates

                                   Standard
Parameter     DF     Estimate         Error    Pr > ChiSq

Intercept      1      -1.9794        0.1646        <.0001
FEMALE         1       0.8430        0.2791        0.0025
```

Women are more likely to have abstained from alcohol than men (p=0.0025), as well as to have drunk $e^{-.068} = .93$ fewer drinks when they drink (p=0.015).

Other significant predictors include substance and age, though model assumptions for count models should always be carefully verified [19].

5.6.4 Negative binomial regression

A negative binomial regression model (5.1.4) might improve on the Poisson.

```
options ls=64;
ods exclude nobs convergencestatus classlevels modelinfo;
proc genmod data=help;
  class substance;
  model i1 = female substance age / dist=negbin;
run;
```

```
The GENMOD Procedure

              Criteria For Assessing Goodness Of Fit

Criterion                     DF           Value        Value/DF

Deviance                     448        539.5954          1.2045
Scaled Deviance              448        539.5954          1.2045
Pearson Chi-Square           448        444.7200          0.9927
Scaled Pearson X2            448        444.7200          0.9927
Log Likelihood                        18884.8073
Full Log Likelihood                   -1708.1668
AIC (smaller is better)                3428.3336
AICC (smaller is better)               3428.5219
BIC (smaller is better)                3453.0290
```

The Deviance / DF is close to 1, suggesting a reasonable fit.

In the following output, the chi-square statistic and the confidence limits for the parameter estimates have been removed.

```
    Analysis Of Maximum Likelihood Parameter Estimates

                                          Standard
Parameter              DF    Estimate        Error    Pr > ChiSq

Intercept               1      1.8681       0.2735        <.0001
FEMALE                  1     -0.2689       0.1272        0.0346
SUBSTANCE    alcohol    1      1.1488       0.1393        <.0001
SUBSTANCE    cocaine    1      0.3252       0.1400        0.0202
SUBSTANCE    heroin     0      0.0000       0.0000         .
AGE                     1      0.0107       0.0075        0.1527
Dispersion              1      1.2345       0.0897

NOTE: The negative binomial dispersion parameter was estimated
      by maximum likelihood.
```

Revisiting the alcohol versus heroin comparison, we see that the effect is still highly significant. The estimated effect is now $e^{1.1488} = 3.15$ times as many drinks for alcohol users—even stronger than in the Poisson model, now that the Poisson requirement that the variance equal the mean has been removed.

5.6.5 Quantile regression

In this section, we fit a quantile regression model (5.3.8) of the number of drinks (`i1`) as a function of predictors, modeling the 75th percentile.

```
ods select parameterestimates;
proc quantreg data=help;
  class substance;
  model i1 = female substance age / quantile=0.75;
run;

The QUANTREG Procedure

             Parameter Estimates

                                  95% Confidence
Parameter         DF Estimate         Limits
Intercept          1   7.0000  -5.3227  16.8570
FEMALE             1  -2.9091  -7.5765   5.1838
SUBSTANCE alcohol  1  22.6364  16.7650  29.6516
SUBSTANCE cocaine  1   3.0909  -3.0808  10.0198
SUBSTANCE heroin   0   0.0000   0.0000   0.0000
AGE                1   0.1818  -0.2154   0.5752
```

The 75th percentile is not significantly affected by gender or age, but is 22.6 drinks higher for alcohol users than heroin users.

5.6.6 Ordinal logit

To demonstrate an ordinal logit analysis (5.1.6), we first create an ordinal categorical variable from the `sexrisk` variable, then model this three-level ordinal variable as a function of `cesd` and `pcs`.

```
data help3;
  set help;
  sexriskcat = (sexrisk ge 2) + (sexrisk ge 6);
run;
```

```
options ls=64; /* make output stay in gray box */
ods select parameterestimates;
proc logistic data=help3 descending;
  model sexriskcat = cesd pcs;
run;

The LOGISTIC Procedure

              Analysis of Maximum Likelihood Estimates

                              Standard          Wald
Parameter     DF   Estimate      Error    Chi-Square    Pr > ChiSq

Intercept 2    1    -0.9436     0.5607        2.8326        0.0924
Intercept 1    1     1.6697     0.5664        8.6909        0.0032
CESD           1   -0.00004    0.00759        0.0000        0.9963
PCS            1    0.00521    0.00881        0.3499        0.5542
```

The descending option makes the lowest sex risk category the reference here, as demonstrated in the next section. Assuming CESD = PCS = 0, the probability that `sexriskcat` is greater than 0 is $\frac{e^{1.6697}}{1+e^{1.6679}} = .84$ (so that the $P(sexriskcat = 0) = .16$, while the probability of the highest category is $\frac{e^{-.9436}}{1+e^{-.9435}} = .28$. From this we can infer that the middle category has a probability of $1-.16-.28 = .56$. The odds ratio for more sex risk increases by $e^{.0051} = 1.005$ with each added PCS unit, though the p-value is large.

In these models, it is important to test the proportional odds assumption. This is part of the default output, its ODS name is `cumulativemodeltest`.

```
options ls=64; /* make output stay in gray box */
ods select cumulativemodeltest;
proc logistic data=help3 descending;
   model sexriskcat = cesd pcs;
run;

The LOGISTIC Procedure

Score Test for the Proportional Odds Assumption

Chi-Square       DF      Pr > ChiSq

    2.0336        2          0.3618
```

We fail to reject the null of proportional odds, and conclude this model is acceptable.

5.6.7 Multinomial logit

We can also fit a multinomial logit (5.1.7) model for the categorized `sexrisk` variable.

```
options ls=64;  /* keep output in grey box */
ods select responseprofile parameterestimates;
proc logistic data=help3 descending;
  model sexriskcat = cesd pcs / link=glogit;
run;
```

```
The LOGISTIC Procedure

           Response Profile

 Ordered                           Total
   Value      sexriskcat       Frequency

       1               2             151
       2               1             244
       3               0              58

Logits modeled use sexriskcat=0 as the reference category.
```

```
              Analysis of Maximum Likelihood Estimates

                                          Standard        Wald
Parameter sexriskcat DF Estimate     Error Chi-Square Pr > ChiSq

Intercept 2           1    0.6863    0.9477      0.5244     0.4690
Intercept 1           1    1.4775    0.8943      2.7292     0.0985
CESD      2           1 -0.00672    0.0132      0.2610     0.6095
CESD      1           1   -0.0133    0.0125      1.1429     0.2850
PCS       2           1    0.0105    0.0149      0.4983     0.4802
PCS       1           1  0.00851    0.0140      0.3670     0.5446
```

Assuming the covariates are 0, the $P(sexriskcat = 2|sexriskcat \neq 1) = \frac{e^{.6863}}{1+e^{.6863}} = .66$. The $P(sexriskcat = 1|sexriskcat \neq 2) = \frac{e^{1.4775}}{1+e^{1.4775}} = .81$. The covariates are still not especially valuable, but the odds of sex risk category 2 vs. 0 go up by $e^{.0105} = 1.01$ per pcs unit, and the odds of sex risk category 1 vs. 0 go up by 1.009 per pcs unit. The similarity of these values, along with their lack of statistical significance, agree with the finding in the previous section that the proportional odds assumption is met.

5.6.8 Generalized additive model

We can fit a generalized additive model (5.3.6), and generate a plot in proc gam using ODS graphics (Figure 5.1).

```
options ls=64; /* stay in gray box */
ods graphics on;
ods select parameterestimates anodev smoothingcomponentplot;
proc gam data=help plots=components(clm);
  class substance;
  model cesd = param(female) spline(pcs) param(substance) /
    method=gcv;
run;
ods graphics off;
```

```
The GAM Procedure
Dependent Variable: CESD
Regression Model Component(s): FEMALE SUBSTANCE
Smoothing Model Component(s): spline(PCS)

                   Regression Model Analysis
                      Parameter Estimates

                      Parameter      Standard
Parameter              Estimate         Error  t Value  Pr > |t|

Intercept              46.35754       2.70426    17.14    <.0001
FEMALE                  4.29244       1.31024     3.28    0.0011
SUBSTANCE alcohol       0.17962       1.36511     0.13    0.8954
SUBSTANCE cocaine      -3.76670       1.43932    -2.62    0.0092
SUBSTANCE heroin              0             .        .         .
Linear(PCS)            -0.27743       0.05278    -5.26    <.0001
```

```
The GAM Procedure
Dependent Variable: CESD
Regression Model Component(s): FEMALE SUBSTANCE
Smoothing Model Component(s): spline(PCS)

                   Smoothing Model Analysis
                     Analysis of Deviance

                                    Sum of
Source                 DF          Squares  Chi-Square  Pr > ChiSq

Spline(PCS)       3.11410      1538.175161     11.3462      0.0111
```

Women have more depressive symptoms than men, and cocaine users fewer than heroin users. There is a strong negative linear association between PCS and CESD ($p < .0001$), but there is also a nonlinear component to this association. The estimated smoothing function on PCS is displayed in Figure 5.1. This shows that for PCS values below about 40 or above about 60, the effect is less negative than the linear effect suggests, while between 40 and 60, the effect is more negative.

5.6.9 Reshaping dataset for longitudinal regression

A wide (multivariate) dataset can be reshaped (2.5.3) into a tall (longitudinal) dataset. Here we create time-varying variables (with a suffix tv) as well as keep baseline values (without the suffix).

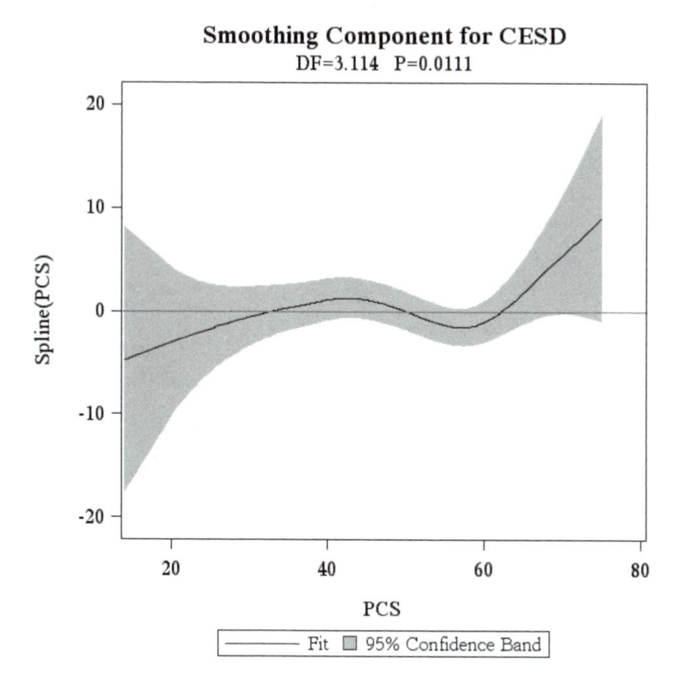

Figure 5.1: Plot of the smoothed relationship between PCS and CESD.

We do this directly with an `output` statement, putting four lines in the `long` dataset for every line in the original dataset.

```
data long;
set help;
  array cesd_a [4] cesd1 - cesd4;
  array mcs_a [4] mcs1 - mcs4;
  array i1_a [4] i11 - i14;
  array g1b_a [4] g1b1 - g1b4;
  do time = 1 to 4;
    cesdtv = cesd_a[time];
    mcstv = mcs_a[time];
    i1tv = i1_a[time];
    g1btv = g1b_a[time];
    output;
  end;
run;
```

We can check the resulting dataset by printing tables by time. In the code below, we use some options to `proc freq` to reduce the information provided by default.

```
proc freq data=long;
  tables g1btv*time / nocum norow nopercent;
run;

The FREQ Procedure

Table of g1btv by time

g1btv     time

Frequency|
Col Pct  |       1|       2|       3|       4|  Total
---------+--------+--------+--------+--------+
       0 |    219 |    187 |    225 |    245 |    876
         |  89.02 |  89.47 |  91.09 |  92.11 |
---------+--------+--------+--------+--------+
       1 |     27 |     22 |     22 |     21 |     92
         |  10.98 |  10.53 |   8.91 |   7.89 |
---------+--------+--------+--------+--------+
Total         246      209      247      266      968

Frequency Missing = 844
```

We can also examine the observations over time for a given individual:

```
proc print data=long;
  where id eq 1;
  var id time cesd cesdtv;
run;

 Obs    ID    time    CESD    cesdtv

 709     1      1      49        7
 710     1      2      49        .
 711     1      3      49        8
 712     1      4      49        5
```

This process can be reversed, creating a wide dataset from a tall one, though this is less commonly necessary.

We begin by using `proc transpose` to make a row for each variable with the four time points in it.

```
proc transpose data=long out=wide1 prefix=time;
by notsorted id;
  var cesdtv mcstv i1tv g1btv;
  id time;
run;
```

Note the `notsorted` option to the `by` statement, which allows us to skip an unneeded `proc sort` step and can be used because we know that all the observations for each `id` are stored adjacent to one another.

This results in the following data.

```
proc print data=wide1 (obs=6);
run;

 Obs    ID    _NAME_      time1       time2       time3       time4

   1     2    cesdtv     11.0000        .           .           .
   2     2    mcstv      41.7270        .           .           .
   3     2    i1tv        8.0000        .           .           .
   4     2    g1btv       0.0000        .           .           .
   5     8    cesdtv     18.0000        .         25.0000       .
   6     8    mcstv      36.0636        .         40.6260       .
```

To put the data for each variable onto one line, we merge the data with itself, taking the lines separately and renaming them along the way using the `where` and `rename` data set options (1.6.1).

```
data wide (drop=_name_);
  merge
  wide1 (where = (_name_="cesdtv")
    rename = (time1=cesd1 time2=cesd2 time3=cesd3 time4=cesd4))
  wide1 (where = (_name_="mcstv")
    rename = (time1=mcs1 time2=mcs2 time3=mcs3 time4=mcs4))
  wide1 (where = (_name_="i1tv")
    rename = (time1=i11 time2=i12 time3=i13 time4=i14))
  wide1 (where = (_name_="g1btv")
    rename = (time1=g1b1 time2=g1b2 time3=g1b3 time4=g1b4));
run;
```

The `merge` without a `by` statement simply places the data from sequential lines in each merged dataset next to each other in the new dataset. Since, here, they are different lines from the same dataset, we know that this is correct. In general, the ability to merge without a by variable can cause unintended consequences.

The final dataset is as desired.

```
proc print data=wide (obs=2);
  var id cesd1 - cesd4;
run;

Obs     ID    cesd1     cesd2     cesd3     cesd4

  1      2      11        .         .         .
  2      8      18        .        25         .
```

5.6.10 Linear model for correlated data

Here we fit a general linear model for correlated data (modeling the covariance matrix directly, 5.2.1).

```
ods select rcorr covparms solutionf tests3;
proc mixed data=long;
  class time;
  model cesdtv = treat time / solution;
  repeated time / subject=id type=un rcorr=7;
run;
```

In this example, the estimated correlation matrix for the 20th subject is printed (this subject was selected because all four time points were observed).

```
The Mixed Procedure

    Estimated R Correlation Matrix for Subject 20

 Row        Col1          Col2          Col3          Col4

   1      1.0000        0.5843        0.6386        0.4737
   2      0.5843        1.0000        0.7430        0.5851
   3      0.6386        0.7430        1.0000        0.7347
   4      0.4737        0.5851        0.7347        1.0000
```

The estimated elements of the variance-covariance matrix are printed row-wise.

```
Covariance Parameter Estimates

Cov Parm      Subject     Estimate

UN(1,1)       ID            207.21
UN(2,1)       ID            125.11
UN(2,2)       ID            221.29
UN(3,1)       ID            131.74
UN(3,2)       ID            158.39
UN(3,3)       ID            205.36
UN(4,1)       ID           97.8055
UN(4,2)       ID            124.85
UN(4,3)       ID            151.03
UN(4,4)       ID            205.75
```

The variance at each time point is around 200, and the covariances are between 100 and 150. Possibly the unstructured correlation matrix is not needed here, and a more parsimonious model might be fit.

```
                       Solution for Fixed Effects

                                     Standard
Effect       time  Estimate             Error     DF    t Value    Pr > |t|

Intercept             21.2439          1.0709    381      19.84      <.0001
TREAT                 -0.4795          1.3196    381      -0.36      0.7165
time          1        2.4140          0.9587    381       2.52      0.0122
time          2        2.6973          0.9150    381       2.95      0.0034
time          3        1.7545          0.6963    381       2.52      0.0121
time          4             0               .      .          .           .
```

The average depression level is smaller at the last time point than at each earlier time.

```
          Type 3 Tests of Fixed Effects

                 Num       Den
Effect            DF        DF     F Value      Pr > F

TREAT              1       381        0.13      0.7165
time               3       381        3.53      0.0150
```

We cannot reject the null hypothesis of no treatment effect, but the joint effect of time is statistically significant.

We can examine data trend and treatment effect in a set of parallel boxplots (Section 6.1.6 and Figure 5.2). Note that the standard boxplot produced by `proc sgpanel` shows the means as diamonds.

```
proc sgpanel data=long;
  panelby time / columns=4;
  vbox cesdtv / category=treat ;
run;
```

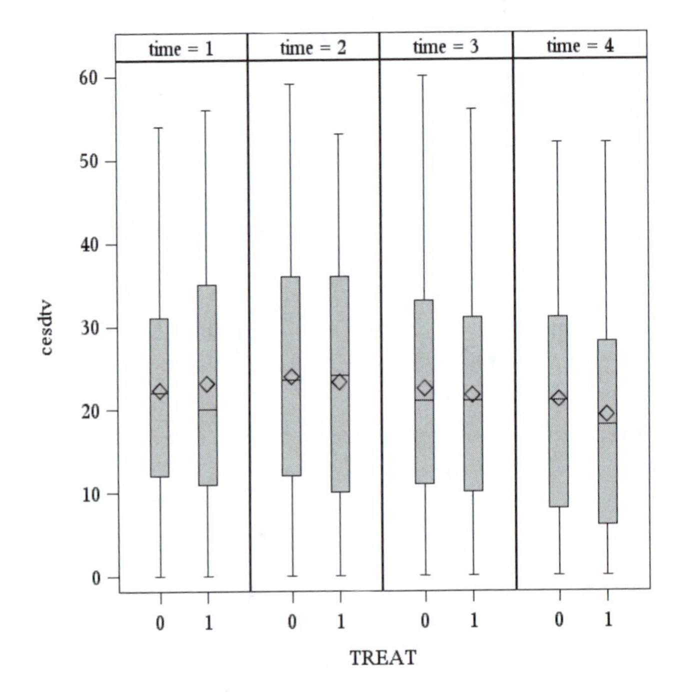

Figure 5.2: Side-by-side boxplots of CESD by treatment and time.

5.6.11 Linear mixed (random slope) model

Here we fix a mixed effects, or random slope model (5.2.3). Note that a given variable can be either a `class` variable or not, within a procedure. In this example, we specify a categorical fixed effect of time but a random slope across time treated continuously. We do this by making a copy of the time variable in a new dataset.

```
data long2;
set long;
  timecopy=time;
run;
```

To make the first time point the referent, we first sort by time; then we use the `order=data` option to the `proc mixed` statement. We save the estimated random effects for later examination, but use `ODS` to suppress their printing.

```
proc sort data= long2; by id descending time; run;

options ls=64;
ods output solutionr=reffs;
ods exclude modelinfo classlevels convergencestatus
  fitstatistics lrt dimensions nobs iterhistory solutionr;
proc mixed data=long2 order=data;
  class timecopy;
  model cesdtv = treat timecopy / solution;
  random int time / subject=id type=un vcorr=20 solution;
run;
```

```
The Mixed Procedure

   Estimated V Correlation Matrix for Subject 20

 Row        Col1         Col2         Col3         Col4

   1      1.0000       0.6878       0.6210       0.5293
   2      0.6878       1.0000       0.6694       0.6166
   3      0.6210       0.6694       1.0000       0.6813
   4      0.5293       0.6166       0.6813       1.0000
```

Observations within individual are strongly correlated over time. Depressive symptoms donot change dramatically differently across people, during treatment.

```
The Mixed Procedure

 Covariance Parameter Estimates

Cov Parm     Subject    Estimate

UN(1,1)      ID           188.43
UN(2,1)      ID         -21.8938
UN(2,2)      ID           9.1731
Residual                 61.5856
```

These are the variances of the random intercepts and slopes, and their covariance.

```
The Mixed Procedure

                    Solution for Fixed Effects

                                      Standard
Effect      timecopy  Estimate           Error    DF  t Value  Pr > |t|

Intercept              23.8843          1.1066   381    21.58    <.0001
TREAT                  -0.4353          1.3333   292    -0.33    0.7443
timecopy    4          -2.5776          0.9438   292    -2.73    0.0067
timecopy    3          -1.0142          0.8689   292    -1.17    0.2441
timecopy    2         -0.06144          0.8371   292    -0.07    0.9415
timecopy    1                0               .     .        .         .
```

As we observed before, depressive symptoms are fewer at time 4 than at time 1. We saw previously that the time 4 depression was smaller than both time 2 and time 3. Here we see that the time 1 depression is similar to time 2 and time 3.

```
The Mixed Procedure

          Type 3 Tests of Fixed Effects

                Num     Den
Effect           DF      DF    F Value    Pr > F

TREAT             1     292       0.11    0.7443
timecopy          3     292       3.35    0.0195
```

As in the previous section, there is little evidence of a treatment effect, and the global effect of time is not null.

To examine the predicted random effects, we can look at the `reffs` dataset created by the ODS `output` statement and the `solution` option to the `random` statement. This dataset includes a `subject` variable created by SAS from the `subject` option in the `random` statement. This contains the same information as the `id` variable, but is encoded as a character variable and has some blank spaces in it. In order to easily print the predicted random effects for the subject with `id=1`, we condition using the `where` statement (1.6.3), removing the blanks using the `strip` function (2.4.9).

```
proc print data=reffs;
  where strip(subject) eq '1';
run;
```

```
                                       StdErr
Obs Effect      Subject Estimate         Pred    DF  tValue  Probt

  1 Intercept      1     -13.4805      7.4764   292   -1.80 0.0724
  2 time           1     -0.02392      2.3267   292   -0.01 0.9918
```

We can check the predicted values for an individual (incorporating their predicted random effect) using the outp option and get the marginal predicted mean from the outpm option to the model statement. Here we suppress all output, then print the observed and predicted values for one subject.

```
ods exclude all;
proc mixed data=long2 order=data;
  class timecopy;
  model cesdtv = treat timecopy / outp=lmmp outpm=lmmpm;
  random int time / subject=id type=un;
run;
ods select all;
```

The lmmp dataset has the predicted mean, conditional on each subject. The lmmpm dataset has the marginal means. If we want to see them in the same dataset, we can merge them (2.5.7). Note that because the input dataset (long2) used in proc mixed was sorted, the output datasets are also sorted. Otherwise, a proc sort step would be needed for each dataset before they could be merged. Since both the datasets contain a variable pred, we rename one of the variables as we merge the datasets.

```
data lmmout;
merge lmmp lmmpm (rename = (pred=margpred));
  by id descending time;
run;

proc print data=lmmout;
  where id eq 1;
  var id time cesdtv pred margpred;
run;

 Obs      ID     time      cesdtv         Pred       margpred

   1       1       4          5        7.29524       20.8714
   2       1       3          8        8.88264       22.4349
   3       1       2          .        9.85929       23.3876
   4       1       1          7        9.94464       23.4490
```

Subject 1 has a large negative predicted random intercept, as we saw in the previous output. This is why the conditional predictions are so much smaller than the marginals. The negative random slope is why the change from time 1 to time 4 is -2.65 for the conditional means and -2.58 for the marginal mean.

5.6.12 Generalized estimating equations (GEEs)

We fit a GEE model (5.2.7), using an unstructured working correlation matrix and empirical variance [26]. To allow for nonmonotone missingness, we use the `within` syntax shown below.

```
ods select geeemppest geewcorr;
proc genmod data=long2 descending;
  class timecopy id;
  model g1btv = treat time / dist=bin;
  repeated subject = id / within=timecopy type=un corrw;
run;
```

The `within` option names a class variable which orders the observations. This is how SAS keeps track of the order of the observations. This is needed for complex correlation structures.

```
The GENMOD Procedure

                Analysis Of GEE Parameter Estimates
                 Empirical Standard Error Estimates

                      Standard   95% Confidence
Parameter Estimate       Error        Limits              Z Pr > |Z|

Intercept   -1.8513     0.2760  -2.3922   -1.3104    -6.71    <.0001
TREAT       -0.0022     0.2683  -0.5280    0.5236    -0.01    0.9935
time        -0.1513     0.0892  -0.3261    0.0236    -1.70    0.0901
```

There is little evidence of treatment effect, but there may be a linear decline across time.

The `corrw` option requests the working correlation matrix be printed.

```
                 Col1          Col2          Col3          Col4

Row1           1.0000        0.3562        0.2671        0.1621
Row2           0.3562        1.0000        0.2801        0.2965
Row3           0.2671        0.2801        1.0000        0.4360
Row4           0.1621        0.2965        0.4360        1.0000
```

Observations closer in time appear to have a greater correlation than observations more distant.

5.6.13 Generalized linear mixed model (GLMM)

Here we fit a GLMM (5.2.6), predicting recent suicidal ideation as a function of treatment, depressive symptoms (CESD), and time. Each subject is assumed to have their own random intercept.

```
options ls=64;  /* make output stay in gray box */
ods select parameterestimates;
proc glimmix data=long;
  model g1btv = treat cesdtv time / dist=bin solution;
  random int / subject=id;
run;

The GLIMMIX Procedure
                        Solutions for Fixed Effects

                          Standard
Effect         Estimate      Error      DF    t Value     Pr > |t|

Intercept       -4.3572     0.4831     381      -9.02       <.0001
TREAT          -0.00749     0.2821     583      -0.03       0.9788
cesdtv          0.07820    0.01027     583       7.62       <.0001
time           -0.09253     0.1111     583      -0.83       0.4051
```

The odds of suicidal ideation increase by a factor of $e^{.0782} = 1.08$ for each unit of `CESD`, for an individual. This is a subject-specific, not a population average effect.

For many generalized linear mixed models, the likelihood has an awkward shape, and maximizing it can be difficult. In such cases, care should be taken to ensure that results are correct. In such settings, it is useful to use numeric integration, rather than the default approximation used by `proc glimmix`; this can be requested using the `method=laplace` option to the `proc glimmix` statement. When results differ, the maximization based on numeric integration of the actual likelihood should be preferred to the analytic iterative maximization of the approximate likelihood.

```
options ls=64;  /* make output stay in gray box */
ods select parameterestimates;
proc glimmix data=long method=laplace;
  model g1btv = treat cesdtv time / dist=bin solution;
  random int / subject=id;
run;
```

```
The GLIMMIX Procedure

                  Solutions for Fixed Effects

                        Standard
Effect       Estimate      Error       DF    t Value    Pr > |t|

Intercept     -8.7633     1.2218      381      -7.17      <.0001
TREAT        -0.04164     0.6707      583      -0.06      0.9505
cesdtv         0.1018    0.01927      583       5.28      <.0001
time          -0.2425     0.1730      583      -1.40      0.1616
```

The p-values did not change materially, but this preferred estimate of the effect of `CESD` is that the odds increase by a factor of 1.11 per CESD unit, 50% greater than in the approximate estimate.

5.6.14 Cox proportional hazards regression model

Here we fit a proportional hazards model (5.3.1) for the time to linkage to primary care, with randomization group, age, gender, and CESD as predictors.

```
options ls=64;
ods exclude modelinfo nobs classlevelinfo convergencestatus
  type3;
proc phreg data=help;
  class treat female;
  model dayslink*linkstatus(0) = treat age female cesd;
run;
```

```
The PHREG Procedure

Summary of the Number of Event and Censored Values

                                           Percent
   Total         Event      Censored      Censored

     431           163           268         62.18
```

```
          Model Fit Statistics

                  Without           With
Criterion      Covariates     Covariates

-2 LOG L         1899.982       1805.368
AIC              1899.982       1813.368
SBC              1899.982       1825.743
```

```
         Testing Global Null Hypothesis: BETA=0

Test                    Chi-Square       DF     Pr > ChiSq

Likelihood Ratio           94.6132        4         <.0001
Score                      92.3599        4         <.0001
Wald                       76.8717        4         <.0001
```

We can reject the null hypothesis that there are no differences in hazard due to the predictors.

```
           Analysis of Maximum Likelihood Estimates

                     Parameter    Standard
Parameter      DF     Estimate       Error  Chi-Square  Pr > ChiSq

TREAT      0    1     -1.65185     0.19324     73.0737      <.0001
AGE             1      0.02467     0.01032      5.7160      0.0168
FEMALE     0    1      0.32535     0.20379      2.5489      0.1104
CESD            1      0.00235     0.00638      0.1363      0.7120

Analysis of Maximum Likelihood Estimates

                   Hazard
Parameter           Ratio

TREAT      0        0.192
AGE                 1.025
FEMALE     0        1.385
CESD                1.002
```

The only predictor with much predictive utility is age; the older people are, the more likely they are to get linked to primary care.

5.6.15 Bayesian Poisson regression

In this example, we fit a Poisson regression model to the count of alcohol drinks in the HELP study as fit previously (5.6.2), this time using Markov Chain Monte Carlo methods.

```
proc import
  datafile='c:/book/help.csv'
  out=help dbms=dlm;
  delimiter=',';
  getnames=yes;
run;
```

```
proc genmod data=help;
  class substance;
  model i1 = female substance age / dist=poisson;
  bayes;
run;
```

The `bayes` statement has options to control many aspects of the MCMC process. While we do not present it here, diagnosis of convergence is a critical part of any MCMC model fitting (see Gelman et al. [16], for an accessible introduction). Diagnostic graphics will be produced if an `ods graphics` statement is submitted. The above code produces the following posterior distribution characteristics for the parameters.

```
                    Posterior Summaries
                                                    Standard
    Parameter                  N           Mean    Deviation
    Intercept              10000         1.7753       0.0579
    female                 10000        -0.1765       0.0279
    substancealcohol       10000         1.1217       0.0336
    substancecocaine       10000         0.3048       0.0378
    age                    10000         0.0132      0.00145

                    Posterior Summaries
                                      Percentiles
     Parameter                25%           50%          75%
     Intercept             1.7357        1.7751       1.8141
     female               -0.1955       -0.1765      -0.1574
     substancealcohol      1.0988        1.1218       1.1447
     substancecocaine      0.2796        0.3046       0.3299
     age                   0.0123        0.0132       0.0142
```

```
                    Posterior Intervals

  Equal-Tail
Parameter         Alpha        Interval              HPD Interval
Intercept         0.050    1.6619     1.8917     1.6612     1.8897
female            0.050   -0.2327    -0.1229    -0.2266    -0.1172
substancealcohol  0.050    1.0555     1.1871     1.0546     1.1861
substancecocaine  0.050    0.2310     0.3801     0.2301     0.3786
age               0.050    0.0104     0.0160     0.0104     0.0160
```

These results are quite similar to the classical model results shown in Section 5.6.2.

5.6.16 Cronbach's α

We calculate Cronbach's α for the 20 items comprising the CESD (Center for Epidemiologic Studies–Depression scale).

```
ods select cronbachalpha;
proc corr data=help alpha nomiss;
  var f1a -- f1t;
run;
ods exclude none;

The CORR Procedure

 Cronbach Coefficient Alpha

Variables                 Alpha
-----------------------------
Raw                    0.760762
Standardized           0.764156
```

The observed α is relatively low: this may be due to ceiling effects for this sample of subjects recruited in a detoxification unit.

5.6.17 Factor analysis

We consider a maximum likelihood factor analysis with varimax rotation for the individual items of the CESD scale. The individual questions can be found in Table A.2. We arbitrarily force three factors.

Before beginning, we exclude observations with missing values.

```
data helpcc;
set help;
  if n(of f1a--f1t) eq 20;
run;

ods select orthrotfactpat factor.rotatedsolution.finalcommunwgt;
proc factor data=helpcc nfactors=3 method=ml rotate=varimax;
  var f1a--f1t;
run;
```

```
The FACTOR Procedure
Rotation Method: Varimax

Final Communality Estimates and Variable Weights
Total Communality: Weighted = 15.332773   Unweighted = 7.811194

Variable    Communality        Weight

F1A          0.25549722    1.34316770
F1B          0.23225517    1.30252990
F1C          0.51565766    2.06467779
F1D          0.29270906    1.41401403
F1E          0.29893385    1.42636367
F1F          0.57894420    2.37499121
F1G          0.23471625    1.30675434
F1H          0.39897919    1.66400037
F1I          0.38389849    1.62312753
F1J          0.37453462    1.59881735
F1K          0.29461104    1.41765736
F1L          0.48551624    1.94346054
F1M          0.11832415    1.13419896
F1N          0.37735132    1.60602564
F1O          0.35641841    1.55382997
F1P          0.59280807    2.45558672
F1Q          0.28734113    1.40315708
F1R          0.53318869    2.14218252
F1S          0.72695038    3.66226205
F1T          0.47255864    1.89596701
```

```
Rotation Method: Varimax

             Factor1           Factor2           Factor3

F1A          0.44823          -0.19780           0.12436
F1B          0.42744          -0.18496           0.12385
F1C          0.61763          -0.29675           0.21479
F1D         -0.25073           0.45456          -0.15236
F1E          0.51814          -0.11387           0.13228
F1F          0.66562          -0.33478           0.15433
F1G          0.47079           0.03520           0.10880
F1H         -0.07422           0.62158          -0.08435
F1I          0.46243          -0.32461           0.25433
F1J          0.49539          -0.22585           0.27949
F1K          0.52291          -0.11535           0.08873
F1L         -0.27558           0.63987           0.01191
F1M          0.28394          -0.03699           0.19061
F1N          0.48453          -0.33040           0.18281
F1O          0.26188          -0.06977           0.53195
F1P         -0.07338           0.75511          -0.13125
F1Q          0.45736          -0.07107           0.27039
F1R          0.61412          -0.28168           0.27696
F1S          0.23592          -0.16627           0.80228
F1T          0.48914          -0.26872           0.40136
```

It is possible to interpret the item scores from the output. We see that the second factor loads on the reverse coded items (H, L, P and D, see 2.13.3). Factor 3 loads on items O and S (*people were unfriendly* and *I felt that people dislike me*).

5.6.18 Linear discriminant analysis

We use linear discriminant analysis to distinguish between homeless and non-homeless subjects, with a prior classification that (by default) half are in each group.

```
ods select lineardiscfunc classifiedresub errorresub;
proc discrim data=help out=ldaout;
  class homeless;
  var age cesd mcs pcs;
run;
```

```
The DISCRIM Procedure

Linear Discriminant Function for HOMELESS

Variable             0             1

Constant     -56.61467     -56.81613
AGE            0.76638       0.78563
CESD           0.86492       0.87231
MCS            0.68105       0.67569
PCS            0.74918       0.73750
```

```
Classification Summary for Calibration Data: WORK.HELP
Resubstitution Summary using Linear Discriminant Function

    From
HOMELESS            0             1         Total

       0          142           102           244
                58.20         41.80        100.00

       1           89           120           209
                42.58         57.42        100.00

   Total          231           222           453
                50.99         49.01        100.00

  Priors          0.5           0.5
```

```
Classification Summary for Calibration Data: WORK.HELP
Resubstitution Summary using Linear Discriminant Function

       Error Count Estimates for HOMELESS

                        0             1         Total

Rate               0.4180        0.4258        0.4219
Priors             0.5000        0.5000
```

The results indicate that homeless subjects tend to be older, have higher CESD scores, and lower MCS and PCS scores.

Figure 5.3 displays the distribution of linear discriminant function values by homeless status; the discrimination ability appears to be slight. The distribution of the linear discriminant function values are shifted to the right for the homeless subjects, though there is considerable overlap between the groups.

```
axis1 label=("Prob(homeless eq 1)");

ods select "Histogram 1";
proc univariate data=ldaout;
  class homeless;
  var _1;
  histogram _1 / nmidpoints=20 haxis=axis1;
run;
```

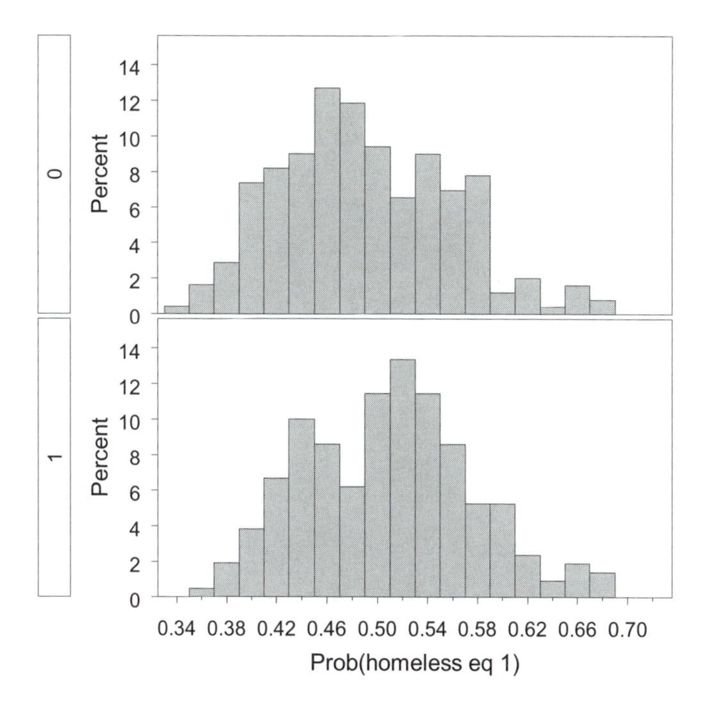

Figure 5.3: Graphical display of assignment probabilities (score functions) from linear discriminant analysis by actual homeless status.

5.6.19 Hierarchical clustering

In this example, we cluster continuous variables from the HELP dataset.

```
ods exclude all;
proc varclus data=help outtree=treedisp centroid;
  var mcs pcs cesd i1 sexrisk;
run;
ods exclude none;

proc tree data=treedisp nclusters=5;
  height _varexp_;
run;
```

Figure 5.4 displays the clustering. Not surprisingly, the MCS and PCS variables cluster together, since they both utilize similar questions and structures. The CESD and I1 variables cluster together, while there is a separate node for SEXRISK.

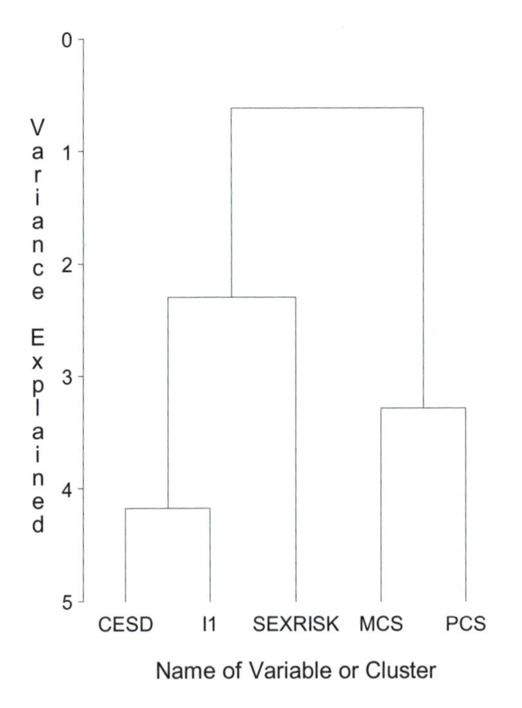

Figure 5.4: Results from hierarchical clustering.

Chapter 6

Graphics

This chapter describes how to create graphical displays, such as scatterplots, boxplots, and histograms. We provide a broad overview of the key ideas and techniques that are available. An additional discussion of ways to annotate displays and change defaults to present publication quality figures is included, as are details regarding how to output graphics in a variety of file formats (Section 6.4). Because graphics are useful to visualize analyses, examples appear throughout the HELP sections at the end of most of the chapters of the book. Graphical tools for exploratory data analysis can also be accessed through one of the point-and-click interfaces (Section 1.9), which include some graphical products that are not easy to replicate using code.

Producing graphics for data analysis is simple and direct. Producing graphics for publication is more complex and typically requires a great deal of time to achieve the desired appearance. Our intent is to provide sufficient guidance that most effects can be achieved, but further investigation of the documentation and experimentation will doubtless be necessary for specific needs. There are a huge number of options: we aim to provide a road map as well as examples to illustrate the power of graphing.

Base SAS supplies character-based plot procedures, but we focus on procedures to create higher-resolution output using SAS/GRAPH. With version 9.2, SAS adds several powerful ways to generate graphics. One is through statements available in existing procedures, as demonstrated in Figure 3.1. Another is `ods graphics` (1.7.3), as demonstrated for example in Figure 5.1. This approach allows graphical output to be produced easily when generating statistical output. Finally, new procedures are introduced in SAS 9.2 which flexibly generate a variety of graphics especially useful in statistical analysis (for an example, see Section 6.6.2).

6.1 A compendium of useful plots

6.1.1 Histogram

Example: See 3.6.1

```
proc univariate data=ds;
  histogram x1 ... xk;
run;
```

or

```
proc gchart data=ds;
  vbar x;
run;
```

The sgplot and sgpanel procedures also generate histograms, but allow fewer options. Including a class statement in proc univariate will result in multiple histograms on the same scale.

The example in Section 3.6.1 demonstrates how to overlay a histogram with a normal or kernel density estimate. Similar estimates are available for the beta, exponential, gamma, lognormal, Weibull, and other densities.

6.1.2 Side-by-side histograms

```
proc gchart data=ds;
  vbar x / group=y type=percent g100;
run;
```

This makes a histogram for each value of y. The type=percent option labels the y axis in percent instead of count, while the g100 option makes each group sum to 100%, rather than the default total bar height summing to 100%.

6.1.3 Barplot

While not typically an efficient graphical display, there are times when a barplot is appropriate to display counts by groups.

```
proc gchart data=ds;
  hbar x1 / sumvar=x2 type=mean;
run; quit;
```

or

```
proc sgplot data=ds;
  hbar x1 / response=x2 stat=mean;
run;
```

The above code produces one bar for each level of X_1 with the length determined by the mean of X_2 in each level. Without the `type=mean` or `stat=mean` option, the length would be the sum of x_2 in each level. With no options, the length of each bar is measured in the number of observations in each level of X_1. The `hbar` statement can be replaced by the `vbar` statement (with identical syntax) to make vertical bars, while the `hbar3d` and `vbar3d` (in `proc gchart` only) make bars with a three-dimensional appearance. Options in `proc gchart` allow display of reference lines, display of statistics, grouping by an additional variable, and many other possibilities. The `sgplot` procedure can also produce similar dot plots using the `dot` statement.

6.1.4 Stem-and-leaf plot

Stem-and-leaf plots are text-based graphics that are particularly useful to describe the distribution of small datasets.

```
proc univariate plot data=ds;
  var x;
run;
```

The stem-and-leaf plot is accompanied by a boxplot; the `plot` option also generates a text-based normal Q-Q plot. To produce only these plots, use an `ods select plots` statement before the `proc univariate` statement.

6.1.5 Boxplot

See also 6.1.6 (side-by-side boxplots) *Example:* See 4.7.6 and 5.6.10

```
data ds2;
  set ds;
  int=1;
run;

proc boxplot data=ds;
  plot x * int;
run;
```

or

```
proc sgplot data=ds;
  vbox x;
run;
```

The `boxplot` procedure is designed to produce side-by-side boxplots (6.1.6). To generate a single boxplot with this procedure, create a variable with the same value for all observations, as above, and make a side-by-side boxplot based on that variable. The sgplot procedure also allows the hbox statement, which produces a horizontal boxplot.

6.1.6 Side-by-side boxplots

See also 6.1.5 (boxplots) *Example:* See 4.7.6 and 5.6.10

```
proc boxplot data=ds;
  plot y * x;
run;
```

or

```
proc boxplot data=ds;
  plot (y1 ... yk) * x (z1 ... zp);
run;
```

or

```
proc sgplot data=ds;
  vbox x / category=y;
run;
```

The first, basic proc boxplot code generates a box describing Y for each level of X. The second, more general proc boxplot code generates a box for each of $Y_1, Y_2, \ldots, Y_k$ for each level of X, further grouped by $Z_1, Z_2, \ldots, Z_p$. The example in Figure 4.6 demonstrates customization.

The proc sgplot code results in boxes of x for each value of y; the similar hbox statement makes horizontal boxplots. The sgpanel procedure can produce multiple side-by-side boxplots in one graphic using vbox or hbox statements similar to those shown for proc sgplot.

6.1.7 Quantile-quantile plot

Example: See 4.7.4

Quantile-quantile plots are a commonly used graphical technique to assess how well a univariate sample of random variables matches a given distribution function. Loosely, the observed quantiles of the data are plotted against the theoretical quantiles of a distribution with parameters estimated from the data. If the data fit the distribution well, a straight line should result.

```
proc univariate data=ds plot;
  var x;
run;
```

or

```
proc univariate data=ds;
  var x;
  qqplot x;
run;
```

The normal Q-Q plot from the `plot` option is a text-based version; it is accompanied by a stem-and-leaf and a boxplot. The plot from the `qqplot` statement is a graphics version, also for the normal distribution. Q-Q plots for other distributions are also available as options to the `qqplot` statement.

6.1.8 Scatterplot

Example: See 4.7.1

See also 6.1.9 (scatterplot with multiple y values) and 6.1.16 (matrix of scatterplots)

```
proc gplot data=ds;
  plot y*x;
run; quit;
```

or

```
proc sgscatter data=ds;
  plot y*x;
run;
```

The `sgpanel` and `sgplot` procedures in SAS 9.2 also generate scatterplots; `proc sgscatter` is particularly useful for scatterplot matrices (6.1.16).

6.1.9 Scatterplot with multiple y values

See also 6.1.16 (matrix of scatterplots) *Example:* See 4.7.2 and 6.6.1

```
proc gplot data=ds; /* create 1 plot with a single y axis */
  plot (y1 ... yk)*x / overlay;
run; quit;
```

or

```
proc gplot data=ds; /* create 1 plot with 2 separate y axes */
  plot y1*x;
  plot2 y2*x;
run; quit;
```

The first code generates a single graphic with all the different Y values plotted. In this case, a simple legend can be added with the `legend` option to the `plot` statement, e.g., `plot (y1 y2)*x / overlay legend`. A fully controllable legend can be added with a `legend` statement as in Figure 2.4.

The second code generates a single graphic with two y-axes. The scale for Y_1 appears on the left and for Y_2 appears on the right.

In either case, the `symbol` statements (see entries in 6.2) can be used to control the plotted values and add interpolated lines as in 6.6.1. SAS will plot each Y value in a different color and/or symbol by default. The `overlay` option and `plot2` statements are not mutually exclusive, so that several variables can be plotted on each Y axis scale.

Using the statement `plot (y1 ... yk)*x` without the `overlay` option will create k separate plots, identical to k separate `proc gplot` procedures. Adding the `uniform` option to the `proc gplot` statement will create k plots with a common y-axis scale.

6.1.10 Bubble plot

A bubble plot is a trivariate plot in which the size of the circle plotted on the scatterplot axes is proportional to a third variable.

```
proc gplot data=ds;
  bubble y*x = z;
run; quit;
```

The circles' areas are proportional to z by default; the radius can be specified instead via the `bscale=radius` option to the `bubble` statement. The areas can be multiplied by a constant with the `bsize` option to the `bubble` statement, which has a default value of 5.

6.1.11 Interaction plots

Example: See 4.7.6

Interaction plots are used to display means by two variables (as in a two-way analysis of variance, 4.1.8).

```
ods graphics on;
  proc glm data=ds;
  class x1 x2;
  model y = x1|x2;
run;
```

In the above, the interaction plot is produced as default output when `ods graphics` are on (1.7.3); the `ods select` statement can be used if only the graphic is desired. In addition, an interaction plot can be generated using the `means` and `gplot` procedures (as shown in 4.7.6).

6.1.12 Conditioning plot

A conditioning plot is used to display a scatterplot for each level of one or two classification variables.

Example: See 6.6.2

```
proc sgpanel data=ds;
  panelby x2 x3;
  scatter x=x1 y=y;
run;
```

A similar plot can be generated with a boxplot, histogram, or other contents in each cell of $X_2 * X_3$ using other `sgplot` statements in place of the `scatter` statement.

6.1.13 Three-dimensional (3-D) plots

Perspective or surface plots, needle plots, and contour plots can be used to visualize data in three dimensions. These are particularly useful when a response is observed over a grid of two-dimensional values.

```
proc g3d data=ds;
  scatter x*y=z;
run;

proc g3d data=ds;
  plot x*y=z;
run;

proc gcontour data=ds;
  plot x*y=z;
run;
```

The `scatter` statement produces a needle plot, a 3-D scatterplot with lines drawn from the points down to the $z = 0$ plane to help visualize the third dimension. The `grid` option to the `scatter` statement may help in clarifying the plot, while the needles can be omitted with the `noneedle` option. The `x` and `y` vars must be a grid for the `plot` statement in either the `g3d` (where it produces a surface plot) or the `gcontour` procedure; if they are not, the `g3grid` procedure can be used to smooth values. The `proc g3d plot` and `scatter` statements accept `rotate` and `tilt` options to show the plot from different perspectives.

6.1.14 Empirical cumulative density function (CDF) plot

```
proc univariate data=ds;
  var x;
  cdfplot x;
run;
```

The empirical density plot offered in `proc univariate` is not smoothed, but theoretical distributions can be superimposed as in the histogram plotted in 3.6.1 and using similar syntax. If a smoothed version is required, it may be necessary to estimate the probability density function (PDF) (6.1.15) and save the output (as shown in 3.6.1), then use it to find the corresponding CDF.

6.1.15 Empirical probability density plot

Example: See 3.6.4, 4.7.4, and 7.3

Density plots are nonparametric estimates of the empirical probability density function.

```
ods graphics on;
  proc kde data=ds;
  univar x1 / plots=(density histdensity);
run;
```

or

```
proc univariate data=ds;
  histogram x / kernel;
run;
```

The `kde` procedure includes kernel density estimation using a normal kernel. The `bivar` statement for `proc kde` will generate a joint empirical density estimate. The bandwidth can be controlled with the `bwm` option and the number of grid points by the `ngrid` option to the `univar` or `bivar` statements. The `out` option to the `univar` or `bivar` statements will save density estimates in a new dataset. The `proc univariate` code generates a graphic (as in 3.1), but no further details.

6.1.16 Matrix of scatterplots

Example: See 6.6.5

```
proc sgscatter data=ds;
  matrix x1 ... xk;
run;
```

The diagonal option to the matrix statement allows the diagonal cells to show, for example, histograms with empirical density estimates. A similar effect can be produced with proc sgpanel as demonstrated in 6.6.5.

6.1.17 Receiver operating characteristic (ROC) curve

Example: See 6.6.4

See also 3.2.2 (diagnostic agreement) and 5.1.1 (logistic regression)

Receiver operating characteristic curves can be used to help determine the optimal cut-score to predict a dichotomous measure. This is particularly useful in assessing diagnostic accuracy in terms of sensitivity (the probability of detecting the disorder if it is present), specificity (the probability that a disorder is not detected if it is not present), and the area under the curve (AUC). The variable x represents a predictor (e.g., individual scores) and y a dichotomous outcome. There is a close connection between the idea of the ROC curve and predictive ability for logistic regression, where the latter allows multiple predictors to be used. ROC curves are embedded in proc logistic.

```
ods graphics on;
proc logistic data=ds plots(only)=roc;
  model y = x1 ... xk;
run;
ods graphics off;
```

The plots(only) option is used to request only the ROC curve be produced, rather than the default inclusion of several additional plots. The probability cutpoint associated with each point on the ROC curve can be printed using roc(id=prob) in place of roc above.

6.1.18 Kaplan–Meier plot

See also 3.4.4 (logrank test) *Example:* See 6.6.3

```
ods graphics on;
ods select survivalplot;
proc lifetest data=ds plots=s;
  time time*status(1);
  strata x;
run;
ods graphics off;
```

or

```
proc lifetest data=ds outsurv=survds;
  time time*status(1);
  strata x;
run;

symbol1 i=stepj r=kx;
proc gplot data=survds;
  plot survival*survtime = x;
run;
```

The second approach demonstrates how to manually construct the plot without using `ods graphics` (1.7.3). The survival estimates generated by `proc lifetest` are saved in a new dataset using the `outsurv` option to the `proc lifetest` statement; we suppose there are `kx` levels of `x`, the stratification variable.

For the plot, a step-function to connect the points is specified using the `i=stepj` option to the `symbol` statement. Finally, `proc gplot` with the `a*b=c` syntax (6.2.2) is called. In this case, `survival*survtime=x` will plot lines for each of the `kx` levels of `x`. Here, `survival` and `survtime` are variable names created by `proc lifetest`. Note that the `r=kx` option to the `symbol` statement is shorthand for typing in the same options for `symbol1`, `symbol2`, ..., `symbolkx` statements; here we repeat them for the `kx` strata specified in `x`.

6.2 Adding elements

Additions to basic plots can be made using a specially formatted dataset called an `annotate` dataset; see Section 7.4.2 for an example. These datasets contain certain required variable names and values. Perfecting a graphic for publication can be facilitated by detailed understanding of `annotate` datasets, a powerful low-level tool. Their use is made somewhat easier by a suite of SAS macros, the `annotate macros` provided with SAS/GRAPH. To use the macros, you must first enable them in the following way.

```
%annomac;
```

You can then call on the macros to draw a line between two points, or plot a circle, and so forth. You do this by creating an `annotate` dataset and calling the macros within it.

```
data annods;
  %system(x, y, s);
  ...
run;
```

Here the ellipses refer to additional `annotate` macros. The `system` macro is useful in getting the macros to work as desired; it defines how the values of x and y in later `annotate` macros are interpreted as well as the size of the plotted values. For example, to measure in terms of the graphics output area, use the value 3 for the first two parameters in the `system` macro. This can be useful for drawing outside the axes. More frequently, we find that using the coordinate system of the plot itself is most convenient; using the value 2 for each parameter will implement this.

6.2.1 Arbitrary straight line

Example: See 4.7.1

```
%annomac;
data annods;
  %system(2,2,2);
  %line(xvalue_1, yvalue_1, xvalue_2, yvalue_2,
    colorspec, linetype, .01);
run;

proc gplot data=ds;
plot x*y / anno=annods;
run; quit;
```

See Section 6.2 for an overview of `annotate` datasets. The `line` macro draws a line from (`xvalue_1`, `yvalue_1`) to (`xvalue_2`, `yvalue_2`). The line will have the color (6.3.11) specified by `colorspec` and be solid or dashed (6.3.9) as specified in `linetype`. The final entry specified the width of the line, here quite narrow. Another approach would be to add the endpoint values to the original dataset, then use the `symbol` statement and the `a*b=c` syntax of `proc gplot` (6.2.2). For the special case of a vertical or horizontal reference line in a scatterplot, the `href=value` or `vref=value` options to the `plot` statement can be used, as in Section 7.5.3.

6.2.2 Plot symbols

Example: See 3.6.2 and 7.5.3

```
symbol1 value=valuename;
symbol1 value='plottext';
symbol1 font=fontname value=plottext;
proc gplot data=ds;
  ...
run;
```

or

```
proc gplot data=ds;
  plot y*x = groupvar;
run; quit;
```

The specific characters to be plotted in `proc gplot` can be controlled using the `value` option to a preceding `symbol` statement as demonstrated in Figure 3.2. The `valuenames` available include `dot`, `point`, `plus`, `diamond`, and `triangle`. They can also be colored with the `color` option and their size changed with the `height` option. The value `none` can be useful if only an interpolated line (6.2.5 and 6.2.6) is desired. A full list of plot symbols can be found in the online help: Contents; SAS Products; SAS/GRAPH; Procedures and Statements; Statements; SYMBOL. The list appears approximately two-thirds of the way through the entry. Additionally, any font character or string can be plotted, if enclosed in quotes as in the second `symbol` statement example, or without the quotes if a `font` option is specified as in the third example.

In the second set of code, a unique plot symbol or color is printed for each value of the variable `group`. If there are many values, for example if `groupvar` is continuous, the results can be confusing.

A troubleshooting tip: If you cannot figure out why a `symbol` statement is not working, try assigning a color, as in 6.3.11.

6.2.3 Add points to an existing graphic

See also 6.2.2 (specifying plotting character) *Example:* See 4.7.1

```
%annomac;
data annods;
  %system(2, 2, 2);
  %circle(xvalue, yvalue, radius);
run;

proc gplot data=ds;
  plot x*y / anno=annods;
run; quit;
```

See Section 6.2 for an introduction to `annotate` datasets. The `circle` macro draws a circle with the center at (`xvalue,yvalue`) and with a radius determined by the last parameter. A suitably small radius will plot a point. Another approach is to add a value to the original dataset, then use the `symbol` statement and the `a*b=c` syntax of `proc gplot` (6.2.2).

6.2.4 Jitter

Example: See 3.6.2

Jittering is the process of adding a negligible amount of noise to each observed value so that the number of observations sharing a value can be easily discerned. This can be accomplished in a data step.

```
data ds;
  set newds;
  jitterx = x + ((uniform(0) * .4) - .2);
run;
```

The above code results in a new dataset with both the original x and its jittered version. The numeric values should be modified to suit the distribution of `x`. These values work acceptably when `x` has a minimum distance between values of 1.

6.2.5 Regression line fit to points

Example: See 4.7.2

```
symbol1 interpol=rl;
proc gplot data=ds;
  plot y*x;
run;
```

or

```
proc sgplot data=ds;
  reg x=x y=y;
run;
```

For `proc gplot`, related interpolations which can be specified in the `symbol` statement are `rq` (quadratic fit) and `rc` (cubic fit). Note also that confidence limits for the mean or for individual predicted values can be plotted by appending `clm` or `cli` after `rx` (see 4.5.6 and 4.5.7). The type of line can be modified as described in 6.3.9. For the `proc sgplot` approach, confidence limits can be requested with the `clm` and/or `cli` options to the `reg` statement; polynomial regression curves can be plotted using the `degree` option. Similar plots can be generated by `proc reg` using `ods graphics` (1.7.3) and by the `sgscatter` and `sgpanel` procedures.

A troubleshooting tip: If you cannot figure out why a `symbol` statement is not working, try assigning a color, as in 6.3.11.

6.2.6 Smoothed line

See also 5.6.8 (generalized additive models) *Example:* See 3.6.2

```
symbol1 interpol=splines;
proc gplot data=ds;
  plot y*x;
run;
```

or

```
ods graphics on;
proc loess data=ds;
  model y = x;
run;
ods graphics off;
```

or

```
ods graphics on;
proc gam data=ds plots=all;
  model y = x;
run;
ods graphics off;
```

or

```
proc sgplot;
  loess x=x y=y;
run;
```

The `spline` interpolation in the `symbol` statement smooths a plot using cubic splines with continuous second derivatives. Other smoothing `interpol` options include `sm`, which uses a cubic spline which minimizes a linear combination of the sum of squares of the residuals and the integral of the square of the second derivative. In that case, an integer between 0 and 99, appended to the `sm` controls the smoothness. Another option is `interpol=lx`, which uses a Lagrange interpolation of degree `x`, where $x = 1, 3, 5$. For all of these smoothers, using the `s` suffix to the method sorts the data internally. If the data are previously sorted, this is not needed. The `sgplot` procedure also offers penalized B-spline smoothing via the `pbspline` statement; the `sgpanel` procedure also includes these smoothers.

6.2.7 Normal density

Example: See 4.7.4

A normal density plot can be added as an annotation to a histogram or empirical density.

```
proc sgplot data=ds;
  density x;
run;
```

or

```
proc univariate data=ds;
  histogram x / normal;
run;
```

In the code above, `sgplot` procedure will draw the estimated normal curve without the histogram. The histogram can be added using the `histogram` statement; the order of these statements determines which element is plotted on top of the other(s). The `univariate` procedure allows many more distributional curves to be fit; it will generate copious text output unless that is suppressed with the `ods select` statement.

6.2.8 Titles

Example: See 3.6.4

```
title 'Title text';
```

or

```
title1 "Main title";
title2 "subtitle";
```

The `title` statement is not limited to graphics, but will also print titles on text output. To prevent any title from appearing after having specified one, use a `title` statement with no quoted title text. Up to 99 numbered `title` statements are allowed. For graphic applications, font characteristics can be specified with options to the `title` statement.

6.2.9 Footnotes

```
footnote 'footnote text';
```

or

```
footnote1 "Main footnote";
footnote2 "subnote";
```

The footnote statement is not limited to graphics, but will also print footnotes on text output. To prevent any footnote from appearing after having specified one, use a footnote statement with no quoted footnote text. Up to 10 numbered footnote statements are allowed. For graphic applications, font characteristics can be specified with options to the footnote statement.

6.2.10 Text

Example: See 3.6.2 and 7.4.2

```
%annomac;
data annods;
  %system(2,2,3);
  %label(xvalue, yvalue, "text", color, angle, rotate, size,
    font, position);
run;

proc gplot data=ds;
  plot x*y / anno=annods;
run; quit;
```

See Section 6.2 for an introduction to annotate datasets. The label macro draws the text text at (xvalue, yvalue), though a character variable can also be specified, if the quotes are omitted. The remainder of the parameters which define the text are generally self-explanatory with the exception of size which is a numeric value measured in terms of the size of the graphics area, and position which specifies the location of the specified point relative to the printed text. A value of 5 centers the text on the specified point. Fonts available include SAS and system fonts; a default typical SAS font is swiss. SAS font information can be found in the online help: Contents; SAS Products; SAS/GRAPH; Concepts; Fonts.

6.2.11 Mathematical symbols

Example: See 2.13.5

Mathematical symbols can be plotted using the text plotting method described in 6.2.10, specifying a font containing math symbols. These can be found in the documentation: Contents; SAS Products; SAS/GRAPH; Concepts; Fonts. Useful fonts include the math and greek fonts. Putting equations with subscripts and superscripts into a plot, or mixing fonts, can be very time-consuming.

6.2.12 Arrows and shapes

Example: See 3.6.4 and 6.6.5

```
%annomac;
data annods;
  %system(2,2,3);
  %arrow(xvalue_1, yvalue_1, xvalue_2, yvalue_2, color,
    linetype, size, angle, font);
  %rect(xvalue_1, yvalue_1, xvalue_2, yvalue_2, color, linetype,
    size);
run;

proc gplot data=ds;
  plot x*y / anno=annods;
run; quit;
```

See Section 6.2 for an introduction to `annotate` datasets. The `arrow` macro draws an arrow from (`xvalue_1`, `yvalue_1`) to (`xvalue_2`, `yvalue_2`). The `size` is a numeric value measured in terms of the size of the graphics area. The `rect` macro draws a rectangle with opposite corners at (`xvalue_1`, `yvalue_1`) and (`xvalue_2`, `yvalue_2`). The type of line drawn is determined by the value of `linetype`, as discussed in 6.3.9, and the color is determined by the value of `color` as discussed in 6.3.11.

6.2.13 Legend

Example: See 2.13.5 and 3.6.4

```
legend1
mode=share
position=(bottom right inside)
across=ncols
frame
label=("Legend Title" h=3)
value=("Grp1" "Grp2");

proc gplot data=ds;
plot y*x=group / legend=legend1;
run;
```

The `legend` statement controls all aspects of how the legend will look and where it will be placed. Legends can be attached to many graphics in a manner similar to that demonstrated above for `proc gplot`. We show some commonly used options. An example of using the `legend` statement can be found in Figure 4.1. The `mode` option determines whether the legend shares the graphic output region with the graphic (shown above); other options reserve space or prevent other plot elements from interfering with the legend. The `position` option places the legend within the plot area. The `across` option specifies the number of columns in the legend. The `frame` option draws a box around the legend. The `label` option describes the text of the legend title, while the `value` option describes the text printed with legend items. Fuller description of the legend statement is provided in the online documentation: Contents; SAS Products; SAS/GRAPH; Procedures and Statements; Statements; LEGEND.

6.2.14 Identifying and locating points

```
symbol1 pointlabel=("#label");
proc gplot data=ds;
  plot y*x;
run;
quit;
```

or

```
data newds;
  set ds;
  label = 'alt=' || x || "," || y;
run;

ods html;
proc gplot data=newds;
  plot y*x / html=label;
run; quit;
ods html close;
```

The first set of code will print the values of the variable `label` on the plot. The variable `label` must appear in the dataset used in the `proc gplot` statement. Note that this can result in messy plots, and it is advisable when there are many observations to choose or create a `label` variable with mostly missing values.

The second set of code will make the value of X and Y appear when the mouse hovers over a plotted data point, as long as the HTML output destination is used. Any text or variable value can be displayed in place of the value of `label`, which in the above entry specifies the observed values of x and y.

6.3 Options and parameters

Many options can be given to plots. In many SAS procedures, these are implemented using `goptions`, `symbol`, `axis`, `legend`, or other statements. Details on these statements can be found in the online help: Contents; SAS Products; SAS/GRAPH; Procedures and Statements; Statements.

6.3.1 Graph size and orientation

```
goptions hsize=Xin vsize=Yin rotate=landscape;
```

or

```
ods graphics width=Xin height=Yin;
```

The size in `goptions` can be specified as above in inches (`in`) or as centimeters (`cm`). Valid `rotate` values also include `portrait`. The size in `ods graphics` (1.7.3) can also be specified as (`cm`), millimeters (`mm`), standard typesetting dimensions (`em`, `en`), or printer's points (`pt`).

6.3.2 Point and text size

Example: See 4.7.6

```
goptions htext=Xin;
title 'titletext' h=Xin;
axis label = ('labeltext' h=Xin);
axis value = ('valuetext' h=Yin);
```

For many graphics statements which produce text, the `h` option controls the size of the printed characters. The default metric is graphic cells, but absolute values in inches and centimeters can also be used as in the `axis` statements shown. The `htext` option to the `goptions` statement affects all text in graphic output unless changed for a specific graphic element.

6.3.3 Box around plots

Example: See 3.6.4

Some graphics-generating statements accept a `frame` or a (default) `noframe` option, which will draw or prevent drawing a box around the plot.

6.3.4 Size of margins

Example: See 4.7.4

The margin options define the printable area of the page for graphics and text.

```
options bottommargin=3in topmargin=4cm
  leftmargin=1 rightmargin=1;
```

The default units are inches; a trailing `cm` indicates centimeters, while a trailing `in` makes inches the explicit metric.

6.3.5 Graphical settings

Example: See 4.7.4

```
goptions reset=all;
```

Many graphical settings are specified using the `goptions` statement. The above usage will revert all values to the SAS defaults.

6.3.6 Multiple plots per page

Example: See 4.7.4

Putting multiple arbitrary plots onto a page is possible but is nontrivial and is beyond the scope of this book. Examples can be found in the online help for `proc greplay`: Contents; SAS Products; SAS Procedures; Proc Greplay. Scatterplot matrices (6.1.16) can be generated using `proc sgscatter` and conditioning plots (6.1.12) can be made using `proc sgpanel`.

6.3.7 Axis range and style

Example: See 4.7.1 and 6.6.1

```
axis1 order = (x1, x2 to x3 by x4, x5);
axis2 order = ("value1" "value2" ... "valuen");
```

Axis statements are associated with vertical or horizontal axes using `vaxis` or `haxis` options in various procedures. Multiple options to the `axis` statement can be listed, as in Figure 4.1. The `axis` statement does not apply to most `ODS graphics` (1.7.3) output.

6.3.8 Axis labels, values, and tick marks

Example: See 2.13.5

```
axis1 label=("Text for axis label" angle=90 color=red
  font=swiss height=2 justify=right rotate=180);
axis1 value=("label1" "label2")
axis1 major=(color=blue height=1.5cm width=2);
axis1 minor=none;
```

Axis statements are associated with vertical or horizontal axes using `vaxis` or `haxis` options in various procedures. For example, in `proc gplot`, one might use a `plot y*x / vaxis=axis1 haxis=axis2` statement. Multiple options to the `axis` statement can be listed, as in Figure 4.1. The `axis` statement does not apply to most `ODS graphics` (1.7.3) output.

In the `label` option above we show the text options available for graphics which apply to both `legend` and `axis` statements, and to `title` statements when graphics are produced. The `angle` option specifies the angle of the line along which the text is printed; the default depends on which `axis` is described. The `color` and `font` options are discussed in Sections 6.3.11 and 6.2.10, respectfully. The `height` option specifies the text size; it is measured in graphic cells, but can be specified with the number of units, for example `height=1cm`. The `justify` option can take values of `left`, `center`, or `right`. The `rotate` option rotates each character in place. The `value` option describes the text which labels the tick marks, and takes the same parameters described for the `label` option.

The `major` and `minor` options take the same parameters; `none` will omit either labeled (`major`) or unlabeled (`minor`) tick marks. The `width` option specifies the thickness of the tick in multiples of the default.

6.3.9 Line styles

Example: See 4.7.1

```
symbol1 interpol=itype line=ltyval;
```

The `interpol` option to the `symbol` statement, which can be shortened to simply `i`, specifies what kind of line should be plotted through the data. Options include smoothers, step functions, linear regressions, and more. The `line` option (which can be shortened to `l`) specifies a solid line (by default, `ltyval=1`) or various dashed or dotted lines (`ltyval 2 ... 33`). A list of line types with associated code can be found in the online documentation: Contents; SAS Products; SAS/GRAPH; Procedures and Statements; Symbol. The line types do not have a separate entry, but appear near the end of the long description of the `symbol` statement.

A troubleshooting tip: If you cannot figure out why a `symbol` statement is not working, try assigning a color, as in 6.3.11.

6.3.10 Line widths

Example: See 2.13.5

```
symbol interpol=interpol_type width=lwdval;
```

When a line through the data is requested using the `interpol` option, the thickness of the line, in multiples of the default thickness, can be specified by the `width` option, for which `w` is a synonym. The default thickness depends on display hardware.

6.3.11 Colors

Example: See 3.6.4

```
symbol1 c=colval cl=colval cv=colval;
axis1 label=(color=colval);
```

Colors can be specified in a variety of ways. Some typical examples of applying colors are shown, but many features of plots can be colored. If precise control is required, `colval` can be specified using a variety of schemes as described in the online documentation: Contents; SAS Products; SAS/GRAPH; Concepts; Colors. For more casual choice of colors, color names such as `blue`, `black`, `red`, `purple`, `strongblue`, or `lightred` can be used.

6.3.12 Log scale

```
axis1 logbase=base logstyle=expand;
```

The `logbase` option scales the axis according to the log of the specified base; valid base values include `e`, `pi`, or a number. The `logstyle` option produces plots with tick marks labeled with the value of the base (`logstyle=power`) or the base raised to that value (`logstyle=expand`).

6.3.13 Omit axes

```
axis1 style=0 major=none minor=none label=("") value=none;
```

To remove an axis entirely, it is necessary to request each element of the axis not be drawn, as shown here.

6.4 Saving graphs

It is straightforward to export graphics in a variety of formats. This can be done using the `ODS` system or via the `goptions` statement. The former will integrate procedure output and graphics. The latter is more cumbersome and cannot be used with `ODS` graphics or the `sgplot`, `sgpanel`, or `sgscatter` procedures. However, it supports more formats, and will work with `gplot`, `gchart`, and other SAS/GRAPH procedures.

6.4.1 PDF

Example: See 7.4.2

```
ods pdf file="filename.pdf";
proc gplot data=ds;
  ...
ods pdf close;
```

or

```
filename filehandle "filename.pdf";
goptions gsfname=filehandle device=pdf gsfmode=replace;

proc gplot data=ds;
  ...
run;
```

In both versions above, the `filename` can include a directory location as well as a name. The `device` option specifies formatting of the graphic; the many valid options can be viewed using `proc gdevice` and key options are presented in this section. The `gsfmode=replace` option allows SAS to create and/or overwrite the graphic. The `filehandle` may not be more than 8 characters long.

The `ods pdf` statement will place graphics and text output from procedures into the pdf file generated.

6.4.2 Postscript

```
ods ps file="filename.ps";
proc gplot data=ds;
  ...
run;
ods ps close;
```

or

```
filename filehandle "filename.ps";
goptions gsfname=filehandle device=ps gsfmode=replace;

proc gplot data=ds;
  ...
run;
```

In both versions above, the `filename` can include a directory location as well as a name. The `device` option specifies formatting of the graphic; the many valid options can be viewed using `proc gdevice` and key options are presented in this section. The `gsfmode=replace` option allows SAS to create and/or overwrite the graphic. The `filehandle` may not be more than 8 characters long.

The `ods ps` statement will place graphics and text output from procedures into the pdf file generated.

6.4.3 RTF

The Rich Text Format (RTF) is a file format developed for cross-platform document sharing. Most word processors are able to read and write RTF documents. The following will create a file in this format containing the graphic; any text generated by procedures will also appear in the RTF file if they are executed between the `ods rtf` and `ods rtf close` statements.

```
ods rtf file="filename.rtf";
proc gplot data=ds;
  ...
run;
ods rtf close;
```

The `filename` can include a directory location as well as a name.

6.4.4 JPEG

```
filename filehandle "filename.jpg";
goptions gsfname=filehandle device=jpeg gsfmode=replace;

proc gplot data=ds;
  ...
run;
```

The `filename` can include a directory location as well as a name. The `device` option specifies formatting of the graphic; valid options can be viewed using `proc gdevice`. The `gsfmode=replace` option allows SAS to create and/or overwrite the graphic. The `filehandle` may not be more than 8 characters long.

6.4.5 Windows Metafile (WMF)

```
filename filehandle "filename.wmf";
goptions gsfname=filehandle device=wmf gsfmode=replace;

proc gplot data=ds;
  ...
run;
```

The `filename` can include a directory location as well as a name. The `device` option specifies formatting of the graphic; valid options can be viewed using `proc gdevice`. The `gsfmode=replace` option allows SAS to create and/or overwrite the graphic. The `filehandle` may not be more than 8 characters long.

6.4.6 Bitmap (BMP)

```
filename filehandle "filename.bmp";
goptions gsfname=filehandle device=bmp gsfmode=replace;

proc gplot data=ds;
  ...
run;
```

The `filename` can include a directory location as well as a name. The `device` option specifies formatting of the graphic; valid options can be viewed using `proc gdevice`. The `gsfmode=replace` option allows SAS to create and/or overwrite the graphic. The `filehandle` may not be more than 8 characters long.

6.4.7 TIFF

```
filename filehandle "filename.tif";
goptions gsfname=filehandle device=tiffp300 gsfmode=replace;

proc gplot data=ds;
  ...
run;
```

The `filename` can include a directory location as well as a name. The `device` option specifies formatting of the graphic; valid options can be viewed using `proc gdevice`. The `gsfmode=replace` option allows SAS to create and/or overwrite the graphic. The `filehandle` may not be more than 8 characters long. Many types of TIFF can be generated; the above `device` specifies a color plot with 300 dpi.

6.4.8 Portable Network Graphic (PNG)

```
filename filehandle "filename.png";
goptions gsfname=filehandle device=png gsfmode=replace;

proc gplot data=ds;
  ...
run;
```

The `filename` can include a directory location as well as a name. The `device` option specifies formatting of the graphic; valid options can be viewed using `proc gdevice`. The `gsfmode=replace` option allows SAS to create and/or overwrite the graphic. The `filehandle` may not be more than 8 characters long.

The `ODS` graphics system works by creating a PNG file which is stored in the current directory or the directory and then creating output in the desired format. So using the `ods output` statement for of the formatted output options in this section will also result in a PNG file.

6.5 Further resources

The books by Tufte [50, 51, 52, 53] provide an excellent framework for graphical displays, some of which builds on the work of Tukey [54].

6.6 HELP examples

To help illustrate the tools presented in this chapter, we apply many of the entries to the HELP data. SAS code can be downloaded from `http://www.math.smith.edu/sas/examples`. We begin by reading in the data.

```
proc import datafile='c:/book/help.csv'
  out=ds dbms=dlm;
  delimiter=',';
  getnames=yes;
run;
```

6.6.1 Scatterplot with multiple axes

The following example creates a single figure that displays the relationship between Center for Epidemiologic Studies–Depression (CESD) and the variables `indtot` (inventory of drug use consequences) and `mcs` (mental component score), for a subset of female alcohol-involved subjects. We specify two different y-axes (6.1.9) for the figure.

```
axis1 minor=none;
axis2 minor=none order=(5 to 60 by 13.625);
axis3 minor=none order=(20, 40, 60);
symbol1 i=sm65s v=circle color=black l=1 w=5;
symbol2 i=sm65s v=triangle color=black l=2 w=5;
proc gplot data=ds;
  where female eq 1 and substance eq 'alcohol';
  plot indtot*cesd / vaxis=axis1 haxis=axis3 legend;
  plot2 mcs*cesd / vaxis = axis2 legend;
run; quit;
```

In the code above, the `symbol` and axis `statements` are used to control the output and to add lines through the data. Note that three axes are specified and are associated with the various axes in the plot in the `vaxis` and `haxis` options to the `plot` and `plot2` statements. The `axis` statements can be omitted for a simpler graphic. The `legend` option produces the simple legend displayed in Figure 6.1. A more attractive legend could be constructed with `legend` statements (one for the `plot` and another for the `plot2` statements) as in Figure 2.4.

6.6.2 Conditioning plot

Figure 6.2 displays a conditioning plot (6.1.12) with the association between MCS and CESD stratified by substance and report of suicidal thoughts (`g1b`).

Note that SAS version 9.2 is required; the plot is hard to replicate with earlier versions of SAS.

```
proc sgpanel data=ds;
  panelby g1b substance / layout=lattice;
  pbspline x=cesd y=mcs;
run; quit;
```

There is a similar association between CESD and MCS for each of the substance groups. Subjects with suicidal thoughts tended to have higher CESD scores, and the association between CESD and MCS was somewhat less pronounced than for those without suicidal thoughts.

6.6.3 Kaplan–Meier plot

The main outcome of the HELP study was time to linkage to primary care, as a function of randomization group. This can be displayed using a Kaplan–Meier plot (see 6.1.18). Detailed information regarding the Kaplan–Meier estimator at each time point can be found by omitting the `ods select` statement. Figure 6.3 displays the estimates, with `+` signs indicating censored observations.

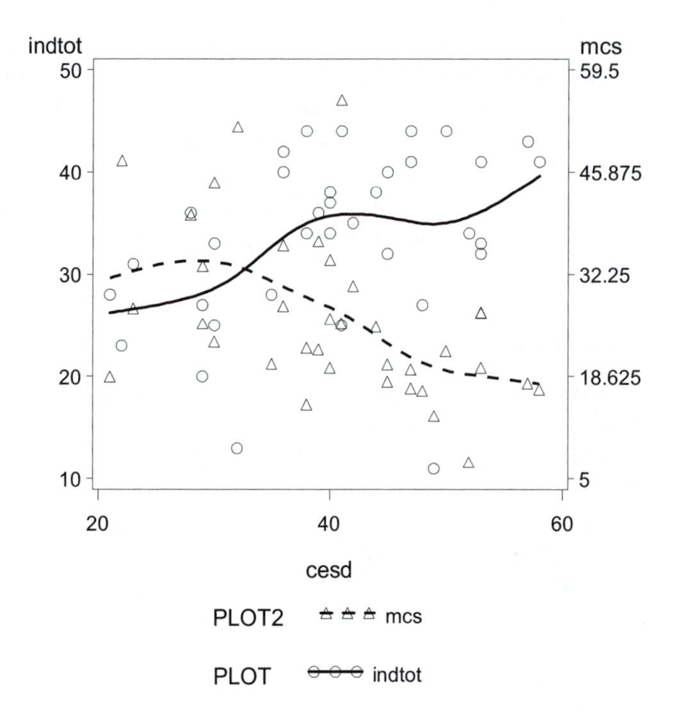

Figure 6.1: Plot of indtot and MCS versus CESD for female alcohol-involved subjects.

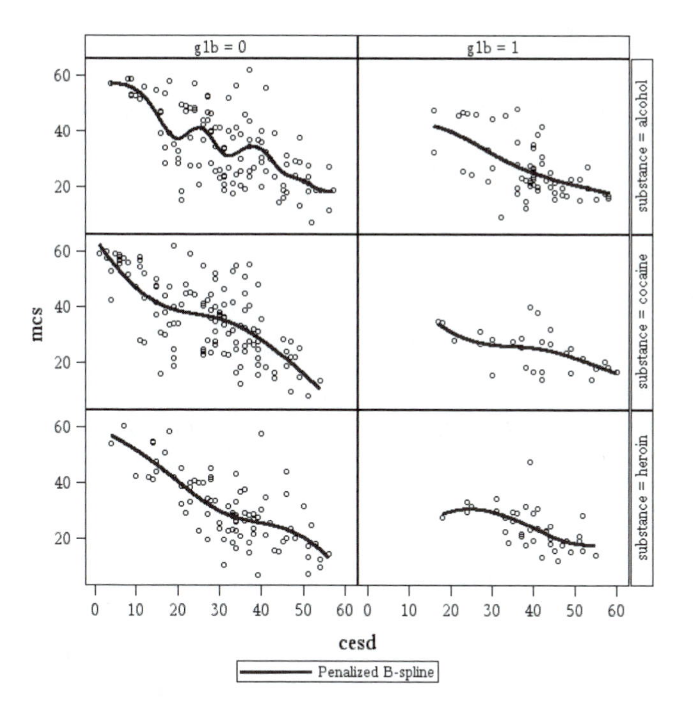

Figure 6.2: Association of MCS and CESD, stratified by substance and report of suicidal thoughts.

```
ods graphics on;
ods select censoredsummary survivalplot;
proc lifetest data=ds plots=s(test);
  time dayslink*linkstatus(0);
  strata treat;
run;
ods graphics off;
```

```
The LIFETEST Procedure

   Summary of the Number of Censored and Uncensored Values

                                                         Percent
Stratum            treat      Total  Failed   Censored  Censored

      1                0        209      35        174     83.25
      2                1        222     128         94     42.34
--------------------------------------------------------------------
  Total                         431     163        268     62.18

NOTE: 22 observations with invalid time, censoring, or strata
values were deleted.
```

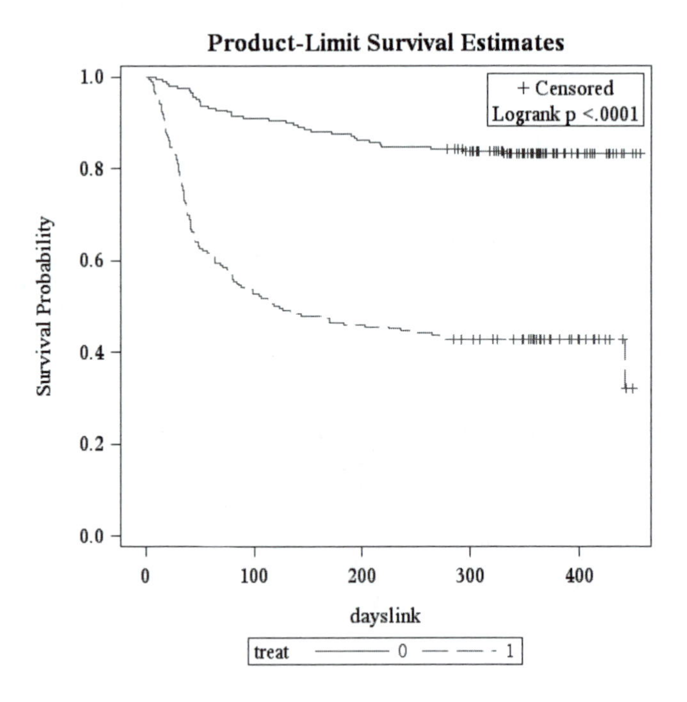

Figure 6.3: Kaplan–Meier estimate of time to linkage to primary care by randomization group.

As reported previously [22, 40], there is a highly statistically significant effect of treatment, with approximately 55% of clinic subjects linking to primary care, as opposed to only 15% of control subjects.

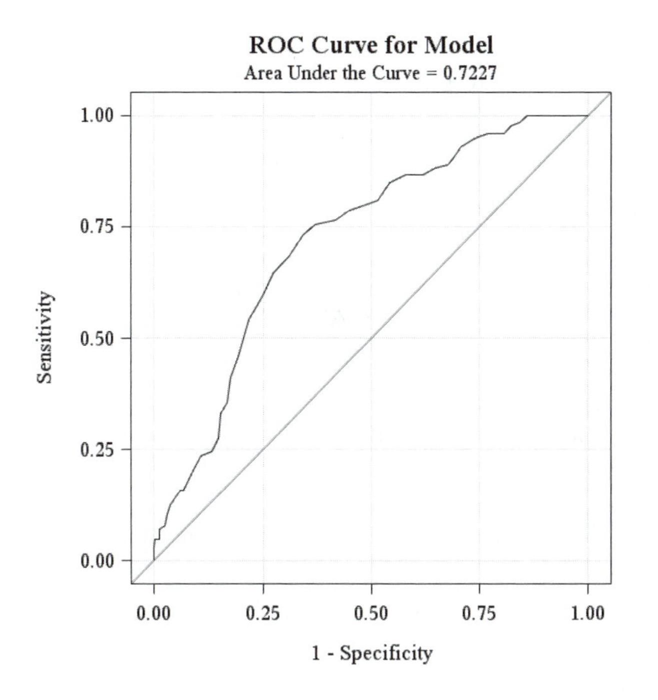

Figure 6.4: Receiver operating characteristic curve for the logistical regression model predicting suicidal thoughts using the CESD as a measure of depressive symptoms (sensitivity = true positive rate; 1-specificity = false positive rate).

6.6.4 ROC curve

Receiver operating characteristic (ROC) curves are used for diagnostic agreement (3.2.2 and 6.1.17) as well as assessing goodness of fit for logistic regression (5.1.1). They can be created using `proc logistic`. Figure 6.4 displays the receiver operating characteristic curve predicting suicidal thoughts using the CESD measure of depressive symptoms.

```
ods graphics on;
ods select roccurve;
proc logistic data=ds descending plots(only)=roc;
  model g1b = cesd;
run;
ods graphics off;
```

The `descending` option changes the behavior of `proc logistic` to model the probability that the outcome is 1; the default models the probability that the outcome is 0.

6.6.5 Pairsplot

We can qualitatively assess the associations between some of the continuous measures of mental health, physical health, and alcohol consumption using a pairsplot or scatterplot matrix (6.1.16). To make the results clearer, we display only the female subjects.

The new `sgscatter` procedure provides a simple way to produce this. The results of the following code are included in Figure 6.5.

```
proc sgscatter data=ds;
  where female eq 1;
  matrix cesd mcs pcs i1 / diagonal=(histogram kernel);
run; quit;
```

If curves through the pairwise scatterplots are required, the following code will produce a similar matrix, with smooth curves in each cell and less helpful graphs in the diagonals (results not shown).

```
proc sgscatter data=ds;
  where female eq 1;
  compare x = (cesd mcs pcs i1)
          y = (cesd mcs pcs i1) / loess;
run; quit;
```

For complete control of the figure, the `sgscatter` procedure will not suffice and more complex coding is necessary; we would begin with SAS macros written by Michael Friendly and available from his Web site at York University.

There is an indication that `CESD`, `MCS`, and `PCS` are interrelated, while `I1` appears to have modest associations with the other variables.

6.6.6 Visualize correlation matrix

One visual analysis which might be helpful to display would be the pairwise correlations. We approximate this in SAS by plotting a confidence ellipse for the observed data. This approach allows an assessment of whether the linear correlation is an appropriate statistic to consider.

In the code below, we demonstrate some options for the `sgscatter` procedure. The `ellipse` option draws confidence ellipses with the requested α-level; here chosen arbitrarily. The `start` option makes the diagonal begin in the lower left; the top left is the default. The `markerattrs` option controls aspects of the appearance of plots generated with the `sgscatter`, `sgpanel`, and `sgplot` procedures.

```
proc sgscatter data=ds;
  matrix mcs pcs pss_fr drugrisk cesd indtot i1 sexrisk /
    ellipse=(alpha=.25) start=bottomleft
    markerattrs=(symbol=circlefilled size=2);
run; quit;
```

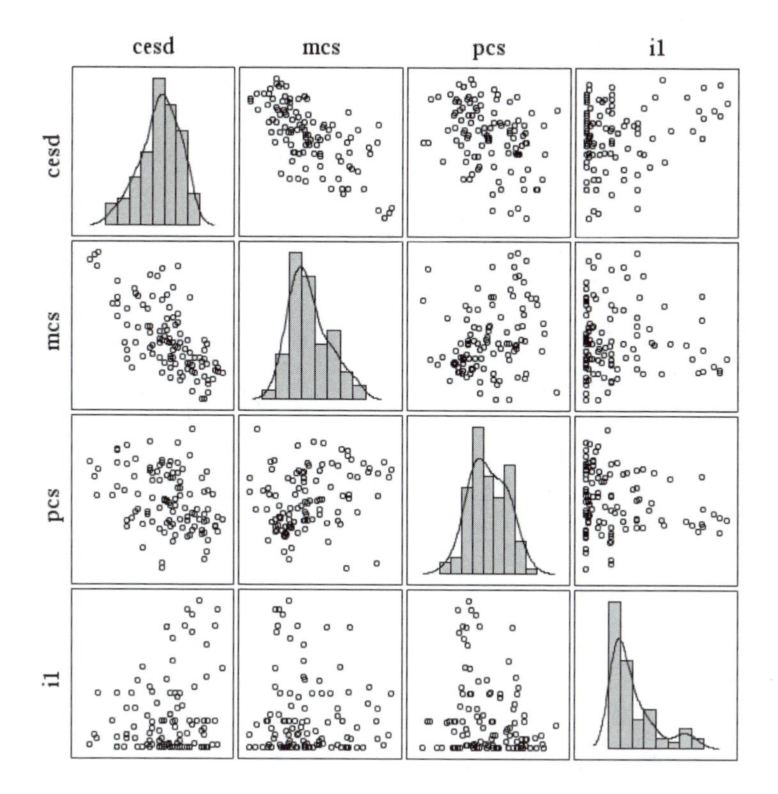

Figure 6.5: Pairsplot of variables from the HELP dataset.

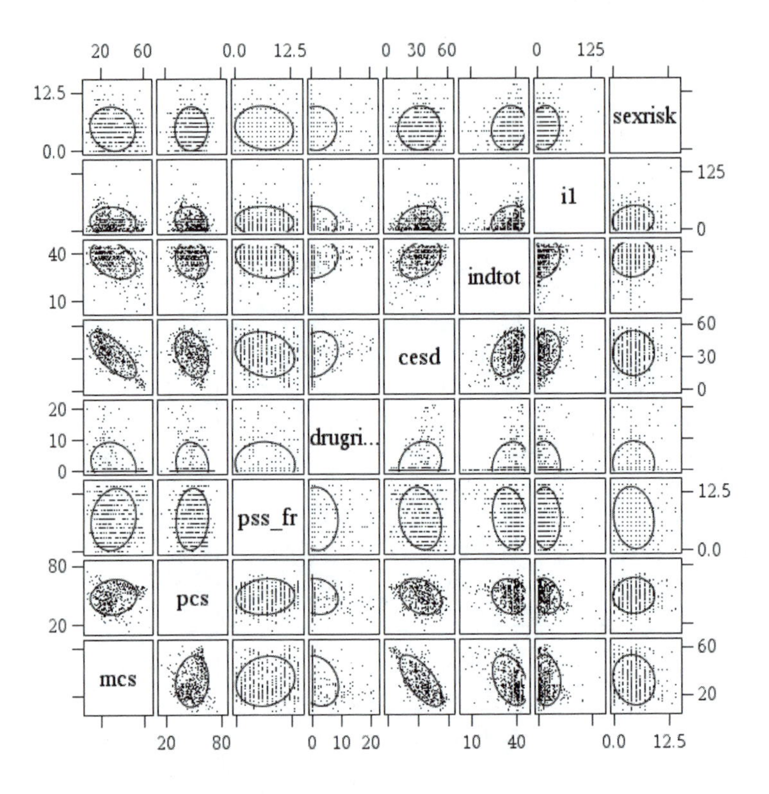

Figure 6.6: Visual display of correlations and associations.

The plot in Figure 6.6 suggests that some of these linear correlations might not be useful measures of association.

Chapter 7

Advanced applications

In this chapter, we address several additional topics that show off the statistical computing strengths and potential of SAS, as well as illustrate many of the entries in the earlier chapters.

7.1 Simulations and data generation

7.1.1 Simulate data from a logistic regression

It is often useful to be able to simulate data from a logistic regression (Section 5.1.1). Our approach is to generate the linear predictor, then apply the inverse link, and finally draw from a distribution with this parameter. This approach is useful in that it can easily be applied to other generalized linear models. In this example we assume an intercept of 0, a single continuous predictor with a slope of 0.5, and generate 1,000 observations. See Section 5.6.1 for an example of fitting logistic regression.

In the following `data` step, we first define parameters for the model and use looping (Section 2.11.1) to replicate the model scenario for random draws of standard normal covariate values (Section 2.10.5), calculating the linear predictor for each, and testing the resulting expit against a random draw from a standard uniform distribution (Section 2.10.3).

```
data testlogistic;
  intercept = 0;
  beta = .5;
  do i = 1 to 1000;
    xtest = normal(12345);
    linpred = intercept + (xtest * beta);
    prob = exp(linpred)/(1 + exp(linpred));
    ytest = uniform(0) lt prob;
    output;
  end;
run;
```

We can fit a logistic regression model and compare the estimated parameter values to the true values which generated the data.

```
options ls=64;
ods select parameterestimates;
proc logistic descending data=testlogistic;
  model ytest = xtest;
run;
ods select all;
```

```
The LOGISTIC Procedure

           Analysis of Maximum Likelihood Estimates

                              Standard          Wald
Parameter    DF    Estimate      Error    Chi-Square    Pr > ChiSq

Intercept     1    -0.00784     0.0649        0.0146        0.9038
xtest         1      0.4515     0.0685       43.4797        <.0001
```

7.1.2 Generate data from generalized linear random effects model

In this example, we generate data from clustered data with a dichotomous outcome. Data of this sort my be modeled with a generalized linear mixed model (5.2.6). In the code below, for 1,500 clusters (denoted by `id`) with 3 observations per cluster are generated. There is a cluster-invariant dichotomous predictor (X_1), a variable indicating order for observations within cluster (X_2) and an additional uniform predictor which varies between observations within cluster (X_3). There are linear effects of X_2 and X_3. Our approach is similar to that used in 7.1.1, except that a random effect is added to each linear predictor.

```
data sim;
  sigbsq=4; beta0=-2; beta1=1.5; beta2=0.5; beta3=-1; n=1500;
  do i = 1 to n;
    x1 = (i lt (n+1)/2);
    randint = normal(0) * sqrt(sigbsq);
    do x2 = 1 to 3 by 1;
      x3 = uniform(0);
      linpred = beta0 + beta1*x1 + beta2*x2 + beta3*x3
              + randint;
      expit = exp(linpred)/(1 + exp(linpred));
      y = (uniform(0) lt expit);
      output;
    end;
  end;
run;
```

This model can be fit using `proc nlmixed` or `proc glimmix`, as shown below. For large datasets like this one, `proc nlmixed` (which uses numerical approximation to integration) can take a prohibitively long time to fit. On the other hand, `proc glimmix` can have trouble converging with the default maximization technique. We show `proc glimmix` options which implement a different maximization technique that may be helpful in such cases.

```
proc nlmixed data=sim qpoints=50;
  parms b0=1 b1=1 b2=1 b3=1;
  eta = b0 + b1*x1 + b2*x2 + b3*x3 + bi1;
  mu = exp(eta)/(1 + exp(eta));
  model y ~ binary(mu);
  random bi1 ~ normal(0, g11) subject=i;
  predict mu out=predmean;
run;
```

or

```
proc glimmix data=sim order=data;
  nloptions maxiter=100 technique=dbldog;
  model y = x1 x2 x3 / solution dist=bin;
  random int / subject=i;
run;
```

7.1.3 Generate binary data with a desired correlation

Correlated dichotomous outcomes Y_1 and Y_2 can be generated using the methods of Lipsitz and colleagues [28]. These are based on a function of the marginal expectations and the desired correlation. Here we generate a sample of 10,000 values where: $P(Y_1 = 1) = .15, P(Y_2 = 1) = .25$ and $\text{Corr}(Y_1, Y_2) = 0.4$, using the **rand** function to generate a multinomial random variate (2.10.4) with the prescribed probabilities corresponding to Y_1, Y_2, both, or neither being set to 1.

```
data test;
  p1=.15; p2=.25; corr=0.4;
  p1p2=corr*sqrt(p1*(1-p1)*p2*(1-p2)) + p1*p2;
  do i = 1 to 10000;
    cat=rand('TABLE', 1-p1-p2+p1p2, p1-p1p2, p2-p1p2);
    y1=0;
    y2=0;
    if cat=2 then y1=1;
    else if cat=3 then y2=1;
    else if cat=4 then do;
      y1=1;
      y2=1;
    end;
    output;
  end;
run;
```

We can check the results with `proc corr` (3.2.3) which conveniently returns the means as well as the correlation. The generated data is close to the desired values.

```
proc corr data=test;
  var y1;
  with y2;
run;

                The CORR Procedure

          1 With Variables:    y2
          1      Variables:    y1

 Variable        N      Mean    Std Dev        Sum    Minimum     Maximum
 y2          10000   0.25470    0.43571       2547          0     1.00000
 y1          10000   0.15290    0.35991       1529          0     1.00000

                                 y1

                   y2       0.41107
                             <.0001
```

7.1.4 Simulate data from a Cox model

To simulate data from a Cox proportional hazards model (5.3.1), we need to model the hazard functions for both time to event and time to censoring. In this example, we use a constant baseline hazard, but this can be modified by specifying other scale parameters for the Weibull random variables.

```
data simcox;
  beta1 = 2;
  beta2 = -1;
  lambdat = 0.002; *baseline hazard;
  lambdac = 0.004; *censoring hazard;
  do i = 1 to 10000;
    x1 = normal(0);
        x2 = normal(0);
        linpred = exp(-beta1*x1 - beta2*x2);
        t = rand("WEIBULL", 1, lambdaT * linpred);
           * time of event;
        c = rand("WEIBULL", 1, lambdaC);
           * time of censoring;
        time = min(t, c);    * which came first?;
        censored = (c lt t);
        output;
  end;
run;
```

This generates data where approximately 40% of the observations are censored. The estimated parameters can be compared to the true values.

```
options ls=64;
ods select censoredsummary parameterestimates;
proc phreg data=simcox;
  model time*censored(1) = x1 x2;
run;
ods select all;
```

```
The PHREG Procedure

Summary of the Number of Event and Censored Values

                                          Percent
   Total          Event    Censored      Censored

   10000           5971        4029         40.29
```

```
The PHREG Procedure

         Analysis of Maximum Likelihood Estimates

                    Parameter    Standard
Parameter   DF       Estimate       Error  Chi-Square  Pr > ChiSq

x1           1        1.98628     0.02213   8059.0716      <.0001
x2           1       -1.01310     0.01583   4098.0277      <.0001

Analysis of Maximum Likelihood Estimates

               Hazard
Parameter       Ratio

x1              7.288
x2              0.363
```

7.2 Power and sample size calculations

Many simple settings lend themselves to analytic power calculations, where closed-form solutions are available. Other situations may require an empirical calculation, where repeated simulation is undertaken.

7.2.1 Analytic power calculation

It is straightforward to find power or sample size (given a desired power) for two-sample comparisons of either continuous or categorical outcomes. We show simple examples for comparing means and proportions in two groups and supply additional information on analytic power calculation available for more complex methods.

```
/* find sample size for two-sample t-test */
proc power;
  twosamplemeans groupmeans=(0 0.5) stddev=1 power=0.9
    ntotal=.;
run;
```

```
/* find power for two-sample t-test */
proc power;
  twosamplemeans groupmeans=(0 0.5) stddev=1 ntotal=200
    power=.;
run;
```

The latter call generates the following output.

```
The POWER Procedure
Two-sample t Test for Mean Difference
     Fixed Scenario Elements
Distribution                       Normal
Method                              Exact
Group 1 Mean                            0
Group 2 Mean                          0.5
Standard Deviation                      1
Total Sample Size                     200
Number of Sides                         2
Null Difference                         0
Alpha                                0.05
Group 1 Weight                          1
Group 2 Weight                          1

Computed Power
Power 0.940
```

```
/* find sample size for two-sample test of proportions */
proc power;
  twosamplefreq test=pchi groupproportions=(.1 .2) power=0.9
    ntotal=.;
run;
```

```
/* find power for two-sample test of proportions */
proc power;
  twosamplefreq test=pchi ntotal=200 groupproportions=(.1 .2)
    power=.;
run;
```

The `power` procedure also allows power calculations for the Wilcoxon rank-sum test, the logrank and related tests for censored data, paired tests of means and proportions, correlations, and for ANOVA and linear and logistic regression.

7.2.2 Simulation-based power calculations

In some settings, analytic power calculations may not be readily available. A straightforward alternative is to estimate power empirically, simulating data under plausible assumptions about the alternative.

We consider a study of children clustered within families. Each family has three children; in some families all three children have an exposure of interest, while in others just one child is exposed. In the simulation, we assume that the outcome is multivariate normal with higher mean for those with the exposure, and 0 for those without. A compound symmetry correlation is assumed, with equal variances at all times. We assess the power to detect an exposure effect where the intended analysis uses a random intercept model (5.2.2) to account for the clustering within families.

With this simple covariance structure it is trivial to generate correlated errors directly, as in the code below; an alternative which could be used for more complex structures would be `proc simnorm` (2.10.6).

```
data simpower1;
  effect = 0.35;   /* effect size */
  corr = 0.4;      /* desired correlation */
  covar = (corr)/(1 - corr);
      /* implied covariance given variance = 1*/
  numsim = 1000;   /* number of datasets to simulate */
  numfams = 100;   /* number of families in each dataset */
  numkids = 3;     /* each family */
  do simnum = 1 to numsim;
    /* make a new dataset for each simnum */
    do famid = 1 to numfams;
    /* make numfams families in each dataset */
      inducecorr = normal(42)* sqrt(covar);
           /* this is the mechanism to achieve the
              desired correlation between kids
  within family */
      do kidnum = 1 to numkids;    /* generate each kid */
        exposed = ((kidnum eq 1) or (famid le numfams/2)) ;
        /* assign kid to be exposed */
        x = (exposed * effect) +
          (inducecorr + normal(0))/sqrt(1 + covar);
        output;
      end;
    end;
  end;
run;
```

In the code above, the integer provided as an argument in the initial use of the `normal` function sets the seed used for all calls to the pseudorandom number generator, so that the results can be exactly replicated, if necessary (see Section 2.10.9).

Next, we fit the intended model to each of the simulated datasets, using the `by` statement (1.6.2). We save the estimated fixed effects parameters (and their standard errors and p-values) using the `ODS` system (1.7). We also suppress all of the output, which would run to thousands of pages.

```
ods select none;
ods output solutionf=simres;
proc mixed data=simpower1 order=data;
by simnum;
  class exposed famid;
  model x = exposed / solution;
  random int / subject=famid;
run;
ods select all;
```

Finally, we process the resulting output dataset to generate an indicator of rejecting the null hypothesis of no exposure effect.

```
data powerout;
set simres;
  where exposed eq 1;  /* This row contains the parameter
                          estimating the effect. */
  reject=(probt lt 0.05);
run;
```

We can then find the proportion of rejections and a confidence interval (3.1.9). This is an empirical estimate of power.

```
proc freq data=powerout;
  tables reject / binomial (level='1');
run;
```

```
The FREQ Procedure
                                         Cumulative    Cumulative
reject    Frequency      Percent        Frequency       Percent
-----------------------------------------------------------------
     0         153        15.30              153          15.30
     1         847        84.70             1000         100.00
```

The confidence limits for the estimated 85% power are important in this setting, though those familiar with analytic power calculation may find it odd.

```
Proportion                 0.8470
ASE                        0.0114
95% Lower Conf Limit       0.8247
95% Upper Conf Limit       0.8693

Exact Conf Limits
95% Lower Conf Limit       0.8232
95% Upper Conf Limit       0.8688
```

We can be 95% confident that the confidence interval $(0.8232, 0.8688)$ holds the true power for this alternative with the stated assumptions.

7.3 Sampling from a pathological distribution

Evans and Rosenthal [13] consider ways to sample from a distribution with density given by:

$$f(y) = c\exp(-y^4)(1 + |y|)^3,$$

where c is a normalizing constant and y is defined on the whole real line. Use of the probability integral transform (Section 2.10.8) is not feasible in this setting, given the complexity of inverting the cumulative density function.

We can find the normalizing constant c using symbolic mathematics software (e.g., Wolfram Alpha, searching for `int(exp(-y^4)(1+y)^3, y=0..infinity)`). This yielded a result of $\frac{1}{4} + \frac{3\sqrt{\pi}}{4} + \Gamma(5/4) + \Gamma(7/4)$ for the integral over the positive real line, which when doubled gives a value of $c = 6.809610784$.

The Metropolis–Hastings algorithm is a Markov Chain Monte Carlo (MCMC) method for obtaining samples from a probability distribution. The premise for this algorithm is that it chooses proposal probabilities so that after the process has converged we are generating draws from the desired distribution. A further discussion can be found in Section 11.3 of *Probability and Statistics: The Science of Uncertainty* [13] or in Section 1.9 of Gelman et al. [16].

We find the acceptance probability $\alpha(x, y)$ in terms of two densities, $f(y)$ and $q(x, y)$ (a proposal density, in our example, normal with specified mean and unit variance) so that

$$\begin{aligned}
\alpha(x,y) &= \min\left\{1, \frac{cf(y)q(y,x)}{cf(x)q(x,y)}\right\} \\
&= \min\left\{1, \frac{c\exp(-y^4)(1+|y|)^3(2\pi)^{-1/2}\exp(-(y-x)^2/2)}{c\exp(-x^4)(1+|x|)^3(2\pi)^{-1/2}\exp(-(x-y)^2/2)}\right\} \\
&= \min\left\{1, \frac{\exp(-y^4+x^4)(1+|y|)^3}{(1+|x|)^3}\right\}
\end{aligned}$$

Begin by picking an arbitrary value for X_1. The Metropolis–Hastings algorithm proceeds by computing the value X_{n+1} as follows:
1. Generate y from a Normal(X_n, 1).
2. Compute $\alpha(x, y)$ as above.
3. With probability $\alpha(x, y)$, let $X_{n+1} = y$ (use proposal value). Otherwise, with probability $1 - \alpha(x, y)$, let $X_{n+1} = X_n = x$ (keep previous value).

The code allows for a burn-in period, which we set at 50,000 iterations, and a desired sample from the target distribution, which we make 5,000. To reduce autocorrelation, we take only every 20th variate.

```
data mh;
  burnin = 50000;
  numvals = 5000;
  thin = 20;
  x = normal(0);
  do i = 1 to (burnin + (numvals * thin));
    y = normal(0) + x;
    switchprob = min(1, exp(-y**4 + x**4) *
      (1 + abs(y))**3 * (1 + abs(x))**(-3));
    if uniform(0) lt switchprob then x = y;
      * if we don't change x, the previous value is retained;
    if (i gt 50000) and mod(i-50000, thin) = 0 then output;
  end;
run;
```

To compare the distribution of the variates with the truth, we first generate a density estimate using `proc kde` (6.1.15), saving the density estimates.

```
ods select none;
proc kde data=mh;
  univar x / out=mhepdf;
run;
ods exclude none;
```

Then we generate the true values, using the constant calculated above.

```
data mh2;
  set mhepdf;
  true = 6.809610784**(-1) *
    exp(-value**4) * (1 + abs(value))**3;
run;
```

Finally, we can plot the estimated and true values together. The results are displayed in Figure 7.1, with the dashed line indicating the true distribution, and the solid line the density estimated from the simulated variates.

```
axis1 label=(angle=90 "Density");
symbol1 i=j l=1 w=3 c=black;
symbol2 i=j l=2 w=3 c=black;
proc gplot data = mh2;
  plot (density true)*value / overlay vaxis = axis1 legend;
  label value="x" density="MH variates";
run;
```

Care is always needed when using MCMC methods. This example was particularly well-behaved, in that the proposal distribution is large compared to the distance between the two modes. Section 6.2 of Lavine [25] and Gelman et al. [16] provide an accessible discussion of these and other issues.

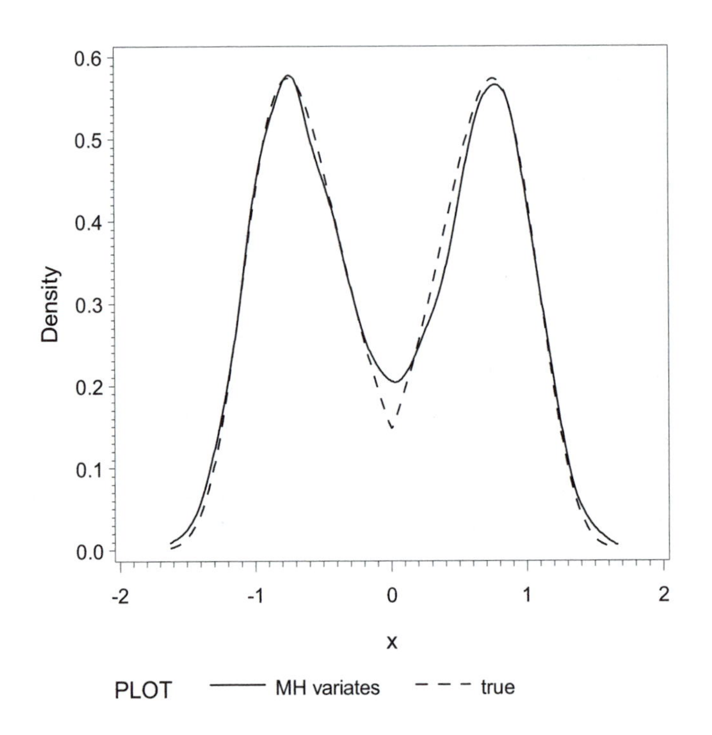

Figure 7.1: Empirical probability density function based on Metropolis–Hastings variates and true density.

7.4 Read variable format files and plot maps

Sometimes datasets are stored in variable format. For example, U.S. Census boundary files (available from `http://www.census.gov/geo/www/cob/index.html`) are available in both proprietary and ASCII formats. An example ASCII file describing the counties of Massachusetts is available on the book Web site. The first few lines are reproduced here.

```
           1       -0.709816806854972E+02        0.427749187746914E+02
      -0.709148990000000E+02        0.428865890000000E+02
      -0.709148860000000E+02        0.428865640000000E+02
      -0.709148860000000E+02        0.428865640000000E+02
      -0.709027680000000E+02        0.428865300000000E+02
      -0.708861360000000E+02        0.428826100000000E+02
      -0.708837340828846E+02        0.428812223551543E+02
...
      -0.709148990000000E+02        0.428865890000000E+02
END
```

The first line contains an identifier for the county (linked with a county name in an additional file) and a longitude and latitude centroid within the polygon representing the county defined by the remaining points. The remaining points on the boundary do not contain the identifier. After the lines with the points, a line containing the word "END" is included. In addition, the county boundaries contain different numbers of points.

7.4.1 Read input files

Reading this kind of data requires some care in programming.

```
filename census1
  url "http://www.math.smith.edu/sas/datasets/co25_d00.dat";

data pcts cents;
  infile census1;
  retain cntyid;
  input @1 endind $3. @;
   /* the trailing '@' means to hold onto this line */
  if endind ne 'END' then do;
    input @7 neglat $1. @;
      /* if this line does not say 'END', then
         check to see if the 7th character is '-' */
    if neglat eq '-' then do;
      /* if so, it has a boundary point */
      input @7 x y;
      output pcts;
     /* write out to boundary dataset */
    end;
    else if neglat ne '-' then do;
      /* if not, it must be the centroid */
      input @9 cntyid 2. x y ;
  output cents;
     /* write it to the centroid dataset */
    end;
  end;
run;
```

Two datasets are defined in the `data` statement, and explicit `output` statements are used to specify which lines are output to which datasets. We begin by reading in the first three characters in each line. The `@` designates the position on the line which is to be read, and in rhw first `input` statement the trailing `@` also "holds" the line for further reading after the end of the `input` statement. If the line is not one which marks the end of a county boundary, we next read in the 7th character. For boundary points, this will be a "-" character, denoting a negative longitude, and we will read the longitude and latitude, then output them to the `pcts` dataset. If the 7th character is not a "-", this must be a centroid, since we already ruled out the possibility that it is an "END" row. If so, we read the county identifier and the coordinates and output the lines to the `cents` dataset.

The county names, which can be associated by the county identifier, are stored in another dataset.

```
filename census2
  url "http://www.math.smith.edu/sas/datasets/co25_d00a.dat";

data cntynames;
infile census2 DSD;
  format cntyname $17. ;
  input cntyid 2. cntyname $;
run;
```

To get the names onto the map, we have to merge the centroid location dataset with the county names dataset. They have to be sorted first.

```
proc sort data=cntynames; by cntyid; run;
proc sort data=cents; by cntyid; run;
```

Note that in the preceding code we depart from the convention of requiring a new line for every statement; simple procedures like these are a convenient place to reduce the line length of the code.

7.4.2 Plotting maps

We are ready to merge the `cntynames` and `cents` datasets. As we merge, we will include the variables needed by the `annotate` facility to put data (the county names) from the merged dataset onto the map. The variables `function`, `style`, `color`, `position`, `when`, `size`, and the `?sys` variables all describe aspects of the text to be placed onto the plot. The text to be plotted must live the variable `text` so we copy the `cntyname` variable into a variable named `text`.

```
data nameloc;
  length function style color $ 8 position $ 1 text $ 20;
  retain xsys ysys "2" hsys "3" when "a";
  merge cntynames cents;
  by cntyid;
  function="label";
  style="swiss";
  text=cntyname;
  color="black";
  size=3;
  position="5";
  output;
run;
```

Finally, we can make the map. The `annotate` option (6.2) tells SAS to use the `nameloc` dataset to mark up the map.

```
ods pdf file="map_plot.pdf";
pattern1 value=empty;
proc gmap map=pcts data=pcts;
  choro const / nolegend coutline=black annotate=nameloc;
  id cntyid;
run; quit;
ods pdf close;
```

The `pattern` statement can be used to control the fill colors when creating chloropleth maps. Here we specify that no fill is needed. Results are displayed in Figure 7.2.

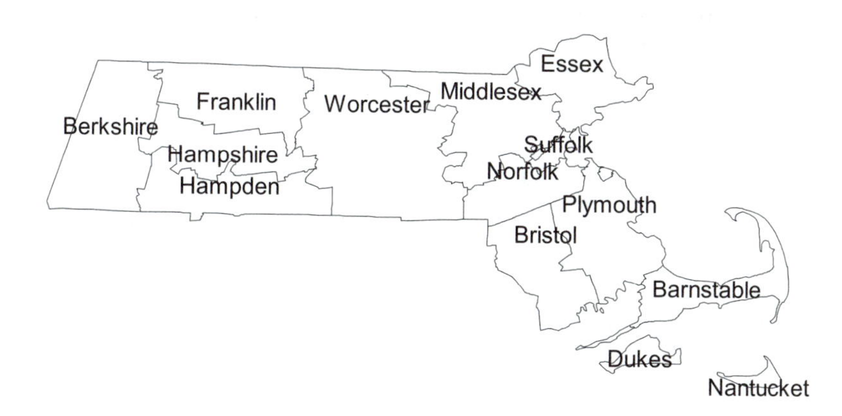

Figure 7.2: Massachusetts counties.

7.5 Data scraping and visualization

In addition to the analytic capabilities available within SAS, the language has the capability of text processing. In the next sections, we demonstrate data harvesting from the Web, by "scraping" a URL, then reading a datafile with two lines per observation, and plotting the results as time series data.

7.5.1 Scraping data from HTML files

As an example, we harvest and display the Amazon bestsellers rank for *The Manga Guide to Statistics* [48]. We can find the bestsellers rank by reading the desired Web page and ferreting out the appropriate line. The following code is highly sensitive to changes if Amazon's page format is changed (but it worked as of January 2010). Note that to avoid printing the long Amazon URL, we created a tinyurl.

We assign the URL an internal name (2.1.6), then input the file using a data step. We exclude all the lines which do not contain the bestsellers rank, using the `count` function (2.4.6). We then extract the rank using the `substr` function (2.4.3), with the `find` function (2.4.6) employed to locate the number within the line. The last step is to turn the extracted text (which contains a comma) into a numeric variable, using formatted input (2.1.3).

```
filename amazon url "http://tinyurl.com/statsmanga";

data test;
infile amazon truncover;
input @1 line $256.;
   if count(line, "Amazon Bestsellers Rank") ne 0;
   rankchar = substr(line, find(line, "#")+1,
        find(line, "in Books") - find(line, "#") - 2);
   rank = input(rankchar, comma9.);
run;
```

7.5.2 Reading data with two lines per observation

We read the bestsellers rank (7.5.1) hourly for several days. We would like to use this data to learn about when and how often a book sells. While a date-stamp was added to the resulting file, unfortunately it was included on a different line. The file (accessible at `http://www.math.smith.edu/sas/data/manga.txt`) has the following form.

```
Thu Dec 31 03:40:03 EST 2009
bestsellers= 30531
Thu Dec 31 04:00:03 EST 2009
bestsellers= 31181
```

We use the infile statement (2.1.3) to read the data, as this affords the most control. As in Section 7.4, we read a key character from the start of the line, holding that line with the trailing "@" character. Then, dependent on the value of that character, we use different input statements to read in the date and time values or the rank. If the line contains the rank, we read the rank, then make a single character value containing the date and time in a format which SAS can interpret. This uses the most recently read values for the date and time variables, even though they appeared on a different line of data. Finally, we convert this character variable into a SAS date-time value (2.1.3 and 2.4.1). Finally, we check the data by printing a few lines.

```
data sales;
infile "c:/book/manga.txt";
retain day month date time edt year;
input @1 type $1 @;
if type ne 's' then do;
  input @1 day $ Month $ date time $ edt $ year;
  end;
else do;
  input @12 rank;
  datetime = compress(date||month||year||"/"||time);
  salestime = input(datetime,datetime18.);
  output;
  end;
run;
```

We check a few observations to be sure they were read correctly.

```
proc print data=sales (obs=6);
   var datetime salestime rank;
run;

Obs         datetime             salestime      rank

  1    30Dec2009/07:14:27      1577776467     18644
  2    30Dec2009/07:20:03      1577776803     18644
  3    30Dec2009/07:40:03      1577778003     18644
  4    30Dec2009/08:00:03      1577779203     18906
  5    30Dec2009/08:20:02      1577780402     18906
  6    30Dec2009/08:40:03      1577781603     18906
```

7.5.3 Plotting time series data

While it is straightforward to plot these data using the proc gplot (6.1.8), we can augment the display by indicating whether the rank was recorded in nighttime (Eastern U.S. time) or not using color (6.3.11) and shape (6.2.2).

We use the `timepart` function (2.6.6) to extract the number of seconds since midnight, and create an indicator of whether the observation happened between 6 P.M. and 8 A.M.

```
data sales2;
set sales;
  if timepart(salestime) lt (8 * 60 * 60) or
     timepart(salestime) gt (18 * 60 * 60) then night = 1;
    else night = 0;
run;
```

The plot is displayed in Figure 7.3, with a vertical line denoting the start of the New Year. The code shows how to use formats to convert date-times into SAS integers on the fly, how to use the software fonts to retrieve a special plotting character (6.2.2), and a vertical reference line (6.2.1). The sales rank gradually increases, presumably between individual book sales, then drops considerably when there is a sale.

```
title;
legend1 mode=share position = (bottom right inside)
  across=1 frame offset=(-12pct)
    label=none value=("Day" "Night");
axis1 minor=none order=
  ("30DEC2009/00:00:00"dt to "11JAN2010/00:00:00"dt by 259200);
axis2 minor=none order=
  (0 to 90000 by 45000) label=(angle=90) value=(angle=90);
symbol1 i=none v=L c=red font=special h = .7;
symbol2 i=none v=dot c=black h = .4;

proc gplot data=sales2;
plot rank * salestime = night/haxis = axis1 vaxis=axis2
  legend=legend1 href="1JAN2010/12:00:00"dt;
  format salestime dtdate5.;
run;
quit;
```

7.6 Missing data: Multiple imputation

Missing data is ubiquitous in most real-world investigations. Here we demonstrate some of the capabilities for fitting incomplete data regression models using multiple imputation [37, 43, 20] implemented with chained equation models [55, 34].

In this example we replicate an analysis from Section 5.6.1 in a version of the HELP dataset that includes missing values for several of the predictors. While not part of the regression model of interest, the `mcs` and `pcs` variables

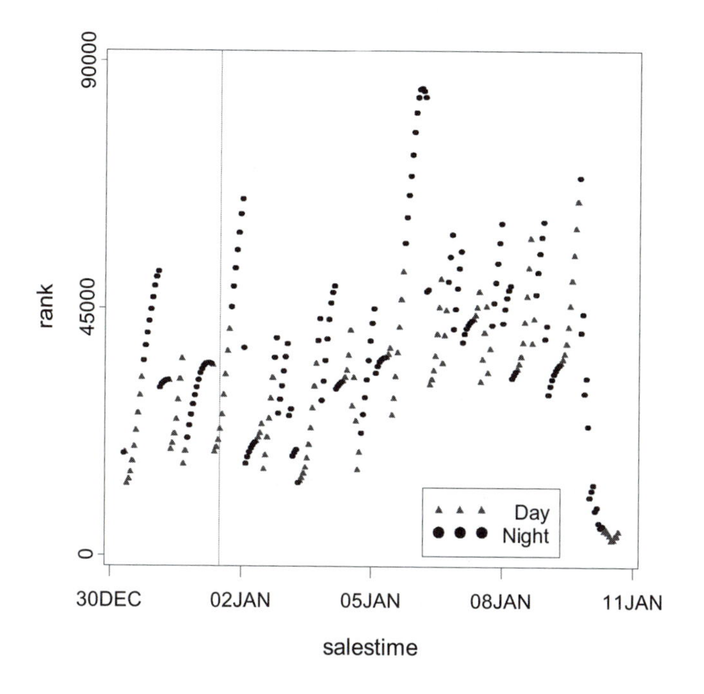

Figure 7.3: Plot of Amazon bestsellers rank over time.

are included in the imputation models, which may make the missing at random assumption more plausible [7].

```
filename myurl
  url "http://www.math.smith.edu/sas/datasets/helpmiss.csv"
  lrecl=704;

proc import replace datafile=myurl out=help dbms=dlm;
  delimiter=',';
  getnames=yes;
run;
```

Running `proc mi` with `nimpute=0` prints a summary of the missing data patterns.

```
ods select misspattern;
proc mi data=help nimpute=0;
  var homeless female i1 sexrisk indtot mcs pcs;
run;
ods select all;
```

```
                          Missing Data Patterns

Group  homeless  female  i1  sexrisk  indtot  mcs  pcs       Freq

    1  X         X       X   X        X       X    X          454
    2  X         X       X   X        X       .    .            2
    3  X         X       X   X        .       X    X           13
    4  X         X       X   .        .       X    X            1
```

```
                        Missing Data Patterns
                   ----------------Group Means---------------
Group    Percent        homeless        female            i1

    1      96.60        0.462555      0.237885     17.920705
    2       0.43        1.000000             0     13.000000
    3       2.77        0.461538      0.230769     31.307692
    4       0.21        1.000000             0     13.000000
```

```
                        Missing Data Patterns
        ----------------------Group Means---------------------
Group         sexrisk          indtot           mcs           pcs

    1        4.638767       35.729075     31.662403     48.018233
    2        7.000000       35.500000             .             .
    3        4.153846               .     27.832265     49.931599
    4               .               .     28.452675     49.938469
```

Since the pattern of missingness is nonmonotone, our options for imputing within SAS are somewhat limited. In the code below, we impute using MCMC. This is not strictly appropriate, since this technique assumes multivariate normal data, which is clearly not the case here. For a summary of multiple imputation options available in SAS, see Horton and Kleinman [20]. An alternative would be to use IVEware, a free suite of SAS macros [35].

```
proc mi data=helpmiss nimpute=20 out=helpmi20 noprint;
  mcmc chain=multiple;
  var homeless female i1 sexrisk indtot mcs pcs;
run;
```

The output dataset `helpmi20` has 20 completed versions of the original dataset, along with an additional variable, `_imputation_`, which identifies the completed versions. We use the by statement in SAS to fit a logistic regression within each completed dataset.

```
ods select none;
ods output parameterestimates=helpmipe covb=helpmicovb;
proc logistic data=helpmi20 descending;
by _imputation_;
  model homeless=female i1 sexrisk indtot / covb;
run;
ods select all;
```

Note the use of the ods select none statement to suppress all printed output and to save the parameter estimates and their estimated covariance matrix for use in multiple imputation.

The multiple imputation inference is performed in proc mianalyze.

```
proc mianalyze parms = helpmipe covb=helpmicovb;
  modeleffects intercept female i1 sexrisk indtot;
run;
```

This generates a fair amount of output; we reproduce only the parameter estimates and their standard errors.

```
                     Parameter Estimates

Parameter        Estimate      Std Error    95% Confidence Limits
intercept       -2.547100       0.596904     -3.71707      -1.37713
female          -0.241332       0.244084     -0.71973       0.23706
i1               0.023101       0.005612      0.01210       0.03410
sexrisk          0.057386       0.035842     -0.01286       0.12763
indtot           0.049641       0.015929      0.01842       0.08086
```

7.7 Further resources

Rubin's review [37] and Schafer's book [43] provide overviews of multiple imputation, while van Buuren, Boshuizen, and Knook [55] and Raghunathan et al. [34] describe chained equation models. Review of software implementations of missing data models can be found in Horton and Lipsitz [21] and Horton and Kleinman [20].

Appendix

The HELP study dataset

A.1 Background on the HELP study

Data from the HELP (Health Evaluation and Linkage to Primary Care) study are used to illustrate many of the entries. The HELP study was a clinical trial for adult inpatients recruited from a detoxification unit. Patients with no primary care physician were randomized to receive a multidisciplinary assessment and a brief motivational intervention or usual care, with the goal of linking them to primary medical care. Funding for the HELP study was provided by the National Institute on Alcohol Abuse and Alcoholism (R01-AA10870, Samet PI) and National Institute on Drug Abuse (R01-DA10019, Samet PI).

Eligible subjects were adults who spoke Spanish or English, reported alcohol, heroin, or cocaine as their first or second drug of choice, and lived close to the primary care clinic to which they would be referred or were homeless. Patients with established primary care relationships they planned to continue, significant dementia, specific plans to leave the Boston area that would prevent research participation, failure to provide contact information for tracking purposes, or pregnancy were excluded.

Subjects were interviewed at baseline during their detoxification stay and follow-up interviews were undertaken every 6 months for 2 years. A variety of continuous, count, discrete, and survival time predictors and outcomes were collected at each of these five occasions.

The details of the randomized trial along with the results from a series of additional analyses have been published [40, 36, 22, 27, 23, 39, 38, 46, 24, 57].

A.2 Road map to analyses of the HELP dataset

Table A.1 summarizes the analyses illustrated using the HELP dataset. These analyses are intended to help illustrate the methods described in the book.

Interested readers are encouraged to review the published data from the HELP study for substantive analyses.

Table A.1: Analyses Undertaken Using the HELP Dataset

Description	Section
Data input and output	2.13.1
Summarize data contents	2.13.1
Data display	2.13.2
Derived variables and data manipulation	2.13.3
Sorting and subsetting	2.13.4
Summary statistics	3.6.1
Bivariate relationship	3.6.2
Contingency tables	3.6.3
Two-sample tests	3.6.4
Survival analysis (logrank test)	3.6.5
Scatterplot with smooth fit	4.7.1
Linear regression with interaction	4.7.3
Regression diagnostics	4.7.4
Fitting stratified regression models	4.7.5
Two-way analysis of variance (ANOVA)	4.7.6
Multiple comparisons	4.7.7
Contrasts	4.7.8
Logistic regression	5.6.1
Poisson regression	5.6.2
Zero-inflated Poisson regression	5.6.3
Negative binomial regression	5.6.4
Quantile regression	5.6.5
Ordinal logit	5.6.6
Multinomial logit	5.6.7
Generalized additive model	5.6.8
Reshaping datasets	5.6.9
General linear model for correlated data	5.6.10
Random effects model	5.6.11
Generalized estimating equations model	5.6.12
Generalized linear mixed model	5.6.13
Proportional hazards regression model	5.6.14
Bayesian Poisson regression	5.6.15
Cronbach α	5.6.16
Factor analysis	5.6.17
Linear discriminant analysis	5.6.18
Hierarchical clustering	5.6.19
Scatterplot with multiple y axes	6.6.1
Conditioning plot	6.6.2

Kaplan–Meier plot	6.6.3
ROC curve	6.6.4
Pairs plot	6.6.5
Visualize correlation matrix	6.6.6
Multiple imputation	7.6

A.3 Detailed description of the dataset

The Institutional Review Board of Boston University Medical Center approved all aspects of the study, including the creation of the de-identified dataset. Additional privacy protection was secured by the issuance of a Certificate of Confidentiality by the Department of Health and Human Services.

A de-identified dataset containing the variables utilized in the end of chapter examples is available for download at the book Web site:
`http://www.math.smith.edu/sas`.

Variables included in the HELP dataset are described in Table A.2. A copy of the study instruments can be found at:
`http://www.math.smith.edu/help`.

Table A.2: Annotated Description of Variables in the HELP Dataset

VARIABLE	DESCRIPTION (VALUES)	NOTE
`a15a`	number of nights in overnight shelter in past 6 months (range 0–180)	see also `homeless`
`a15b`	number of nights on the street in past 6 months (range 0–180)	see also `homeless`
`age`	age at baseline (in years) (range 19–60)	
`anysubstatus`	use of any substance post-detox (0=no, 1=yes)	see also `daysanysub`
`cesd`*	Center for Epidemiologic Studies Depression scale (range 0–60)	see also `f1a`–`f1t`
`d1`	how many times hospitalized for medical problems (lifetime) (range 0–100)	
`daysanysub`	time (in days) to first use of any substance post-detox (range 0–268)	see also `anysubstatus`

`daysdrink`	time (in days) to first alcoholic drink post-detox (range 0–270)	see also `drinkstatus`
`dayslink`	time (in days) to linkage to primary care (range 0–456)	see also `linkstatus`
`drinkstatus`	use of alcohol post-detox (0=no, 1=yes)	see also `daysdrink`
`drugrisk`*	Risk-Assessment Battery (RAB) drug risk score (range 0–21)	see also `sexrisk`
`e2b`*	number of times in past 6 months entered a detox program (range 1–21)	
`f1a`	I was bothered by things that usually don't bother me (range 0–3#)	
`f1b`	I did not feel like eating; my appetite was poor (range 0–3#)	
`f1c`	I felt that I could not shake off the blues even with help from my family or friends (range 0–3#)	
`f1d`	I felt that I was just as good as other people (range 0–3#)	
`f1e`	I had trouble keeping my mind on what I was doing (range 0–3#)	
`f1f`	I felt depressed (range 0–3#)	
`f1g`	I felt that everything I did was an effort (range 0–3#)	
`f1h`	I felt hopeful about the future (range 0–3#)	
`f1i`	I thought my life had been a failure (range 0–3#)	
`f1j`	I felt fearful (range 0–3#)	
`f1k`	My sleep was restless (range 0–3#)	
`f1l`	I was happy (range 0–3#)	
`f1m`	I talked less than usual (range 0–3#)	
`f1n`	I felt lonely (range 0–3#)	
`f1o`	People were unfriendly (range 0–3#)	
`f1p`	I enjoyed life (range 0–3#)	

f1q	I had crying spells (range 0–3#)	
f1r	I felt sad (range 0–3#)	
f1s	I felt that people dislike me (range 0–3#)	
f1t	I could not get going (range 0–3#)	
female	gender of respondent (0=male, 1=female)	
g1b*	experienced serious thoughts of suicide (last 30 days, values 0=no, 1=yes)	
homeless*	1 or more nights on the street or shelter in past 6 months (0=no, 1=yes)	see also a15a and a15b
i1*	average number of drinks (standard units) consumed per day (in the past 30 days, range 0–142)	see also i2
i2	maximum number of drinks (standard units) consumed per day (in the past 30 days range 0–184)	see also i1
id	random subject identifier (range 1–470)	
indtot*	Inventory of Drug Use Consequences (InDUC) total score (range 4–45)	
linkstatus	post-detox linkage to primary care (0=no, 1=yes)	see also dayslink
mcs*	SF-36 Mental Component Score (range 7–62)	see also pcs
pcrec*	number of primary care visits in past 6 months (range 0–2)	see also linkstatus, not observed at baseline
pcs*	SF-36 Physical Component Score (range 14–75)	see also mcs
pss_fr	perceived social supports (friends, range 0–14)	see also dayslink
satreat	any BSAS substance abuse treatment at baseline (0=no, 1=yes)	

`sexrisk`*	Risk-Assessment Battery (RAB) drug risk score (range 0–21)	see also `drugrisk`
`substance`	primary substance of abuse (alcohol, cocaine, or heroin)	
`treat`	randomization group (0=usual care, 1=HELP clinic)	

Notes: Observed range is provided (at baseline) for continuous variables.
* denotes variables measured at baseline and follow-up (e.g., `cesd` is baseline measure, `cesd1` is measure at 6 months, and `cesd4` is measure at 24 months).
#: For each of the 20 items in HELP section F1 (CESD), respondents were asked to indicate how often they behaved this way during the past week (0 = rarely or none of the time, less than 1 day; 1 = some or a little of the time, 1 to 2 days; 2 = occasionally or a moderate amount of time, 3 to 4 days; or 3 = most or all of the time, 5 to 7 days); items `f1d`, `f1h`, `f1l` and `f1p` were reverse-coded.

Bibliography

[1] D. Adams. *The Hitchhiker's Guide to the Galaxy.* London, UK: Pan Books, 1979.

[2] A. Agresti. *Categorical Data Analysis.* New York: John Wiley & Sons, 2002.

[3] A. H. Bowker. Bowker's test for symmetry. *Journal of the American Statistical Association*, 43:572–574, 1948.

[4] R. P. Cody and J. K. Smith. 4th ed. *Applied Statistics and the SAS Programming Language.* Prentice Hall, 1997.

[5] D. Collett. *Modelling Binary Data.* New York: Chapman & Hall, 1991.

[6] D. Collett. *Modeling Survival Data in Medical Research.* 2nd ed. Boca Raton, FL: Chapman & Hall/CRC Press, 2003.

[7] L. M. Collins, J. L. Schafer, and C.-M. Kam. A comparison of inclusive and restrictive strategies in modern missing data procedures. *Psychological Methods*, 6(4):330–351, 2001.

[8] R. D. Cook. *Residuals and Influence in Regression.* New York: Chapman & Hall, 1982.

[9] L. D. Delwiche and S. J. Slaughter. *The Little SAS Book: A Primer.* 3rd ed. Cary, NC: SAS Publishing, 2003.

[10] A. J. Dobson and A. Barnett. *An Introduction to Generalized Linear Models.* 3rd ed. Boca Raton, FL: Chapman & Hall/CRC Press, 2008.

[11] M. Dwass. Modified randomization tests for nonparametric hypotheses. *Annals of Mathematical Statistics*, 28:181–187, 1957.

[12] B. Efron and R. J. Tibshirani. *An Introduction to the Bootstrap.* New York: Chapman & Hall, 1993.

[13] M. J. Evans and J. S. Rosenthal. *Probability and Statistics: The Science of Uncertainty.* New York: W. H. Freeman and Company, 2004.

[14] G. M. Fitzmaurice, N. M. Laird, and J. H. Ware. *Applied Longitudinal Analysis.* New York: Wiley, 2004.

[15] T. R. Fleming and D. P. Harrington. *Counting Processes and Survival Analysis.* New York: John Wiley & Sons, 1991.

[16] A. Gelman, J. B. Carlin, H. S. Stern, and D. B. Rubin. *Bayesian Data Analysis.* 2nd ed. Boca Raton, FL: Chapman & Hall/CRC Press, 2004, p. 25.

[17] P. I. Good. *Permutation Tests: A Practical Guide to Resampling Methods for Testing Hypotheses.* Springer-Verlag, 1994.

[18] J. W. Hardin and J. M. Hilbe. *Generalized Estimating Equations.* Boca Raton: FL: Chapman & Hall/CRC Press, 2002.

[19] N. J. Horton, E. Kim, and R. Saitz. A cautionary note regarding count models of alcohol consumption in randomized controlled trials. *BMC Medical Research Methodology,* 7(9), 2007.

[20] N. J. Horton and K. P. Kleinman. Much ado about nothing: A comparison of missing data methods and software to fit incomplete data regression models. *The American Statistician,* 61:79–90, 2007.

[21] N. J. Horton and S. R. Lipsitz. Multiple imputation in practice: Comparison of software packages for regression models with missing variables. *The American Statistician,* 55(3):244–254, 2001.

[22] N. J. Horton, R. Saitz, N. M. Laird, and J. H. Samet. A method for modeling utilization data from multiple sources: Application in a study of linkage to primary care. *Health Services and Outcomes Research Methodology,* 3:211–223, 2002.

[23] S. G. Kertesz, N. J. Horton, P. D. Friedmann, R. Saitz, and J. H. Samet. Slowing the revolving door: Stabilization programs reduce homeless persons' substance use after detoxification. *Journal of Substance Abuse Treatment,* 24:197–207, 2003.

[24] M. J. Larson, R. Saitz, N. J. Horton, C. Lloyd-Travaglini, and J. H. Samet. Emergency department and hospital utilization among alcohol- and drug-dependent detoxification patients without primary medical care. *American Journal of Drug and Alcohol Abuse,* 32:435–452, 2006.

[25] M. Lavine. *Introduction to Statistical Thought.* http://*www.math.umass.edu/~lavine/Book/book.html.* 2005.

[26] K. -Y. Liang and S. L. Zeger. Longitudinal data analysis using generalized linear models. *Biometrika,* 73:13–22, 1986.

[27] J. Liebschutz, J. B. Savetsky, R. Saitz, N. J. Horton, C. Lloyd-Travaglini, and J. H. Samet. The relationship between sexual and physical abuse and substance abuse consequences. *Journal of Substance Abuse Treatment*, 22(3):121–128, 2002.

[28] S. R. Lipsitz, N. M. Laird, and D. P. Harrington. Maximum likelihood regression methods for paired binary data. *Statistics in Medicine*, 9:1517–1525, 1990.

[29] R. Littell, W. W. Stroup, and R. Freund. *SAS For Linear Models.* 4th ed. Cary, NC: SAS Publishing, 2002.

[30] B. F. J. Manly. *Multivariate Statistical Methods: A Primer.* 3rd ed. Boca Raton, FL: Chapman & Hall/CRC Press, 2004.

[31] P. McCullagh and J. A. Nelder. *Generalized Linear Models.* New York: Chapman & Hall, 1989.

[32] P. Murrell. *Introduction to Data Technologies.* Boca Raton, FL: Chapman & Hall/CRC Press, 2009.

[33] National Institute of Alcohol Abuse and Alcoholism, Bethesda, MD. *Helping Patients Who Drink Too Much*, 2005.

[34] T. E. Raghunathan, J. M. Lepkowski, J. van Hoewyk, and P. Solenberger. A multivariate technique for multiply imputing missing values using a sequence of regression models. *Survey Methodology*, 27(1):85–95, 2001.

[35] T. E. Raghunathan, P. W. Solenberger, and J. V. Hoewyk. IVEware: Imputation and variance estimation software. *http://www.isr.umich.edu/src/smp/ive* (accessed August 10, 2006).

[36] V. W. Rees, R. Saitz, N. J. Horton, and J. H. Samet. Association of alcohol consumption with HIV sex- and drug-risk behaviors among drug users. *Journal of Substance Abuse Treatment*, 21(3):129–134, 2001.

[37] D. B. Rubin. Multiple imputation after 18+ years. *Journal of the American Statistical Association*, 91:473–489, 1996.

[38] R. Saitz, N. J. Horton, M. J. Larson, M. Winter, and J. H. Samet. Primary medical care and reductions in addiction severity: A prospective cohort study. *Addiction*, 100(1):70–78, 2005.

[39] R. Saitz, M. J. Larson, N. J. Horton, M. Winter, and J. H. Samet. Linkage with primary medical care in a prospective cohort of adults with addictions in inpatient detoxification: Room for improvement. *Health Services Research*, 39(3):587–606, 2004.

[40] J. H. Samet, M. J. Larson, N. J. Horton, K. Doyle, M. Winter, and R. Saitz. Linking alcohol and drug dependent adults to primary medical care: A randomized controlled trial of a multidisciplinary health intervention in a detoxification unit. *Addiction*, 98(4):509–516, 2003.

[41] J. -M. Sarabia, E. Castillo, and D. J. Slottje. An ordered family of Lorenz curves. *Journal of Econometrics*, 91:43–60, 1999.

[42] C. E. Särndal, B. Swensson, and J. Wretman. *Model Assisted Survey Sampling*. New York: Springer-Verlag, 1992.

[43] J. L. Schafer. *Analysis of Incomplete Multivariate Data*. New York: Chapman & Hall, 1997.

[44] R. L. Schwart, T. Phoenix, and B. D. Foy. *Learning Perl*. 5th ed. Sebastopol, CA: O'Reilly and Associates, 2008.

[45] G. A. F. Seber and C. J. Wild. *Nonlinear Regression*. New York: John & Wiley Sons, 1989.

[46] C. W. Shanahan, A. Lincoln, N. J. Horton, R. Saitz, M. J. Larson, and J. H. Samet. Relationship of depressive symptoms and mental health functioning to repeat detoxification. *Journal of Substance Abuse Treatment*, 29:117–123, 2005.

[47] B. G. Tabachnick and L. S. Fidell. *Using Multivariate Statistics*. 5th ed. Boston, MA: Allyn & Bacon, 2007.

[48] S. Takahashi. *The Manga Guide to Statistics*. San Francisco, CA: No Starch Press, 2008.

[49] R. Tibshirani. Regression shrinkage and selection via the lasso. *Journal of the Royal Statistical Society B*, 58(1), 1996.

[50] E. R. Tufte. *Envisioning Information*. Cheshire, CT: Graphics Press, 1990.

[51] E. R. Tufte. *Visual Explanations: Images and Quantities, Evidence and Narrative*. Cheshire, CT: Graphics Press, 1997.

[52] E. R. Tufte. *Visual Display of Quantitative Information*. 2nd ed. Cheshire, CT: Graphics Press, 2001.

[53] E. R. Tufte. *Beautiful Evidence*. Cheshire, CT: Graphics Press, 2006.

[54] J. W. Tukey. *Exploratory Data Analysis*. Reading, MA: Addison Wesley, 1977.

[55] S. van Buuren, H. C. Boshuizen, and D. L. Knook. Multiple imputation of missing blood pressure covariates in survival analysis. *Statistics in Medicine*, 18:681–694, 1999.

[56] B. West, K. B. Welch, and A. T. Galecki. *Linear Mixed Models: A Practical Guide Using Statistical Software*. Boca Raton, FL: Chapman & Hall/CRC Press, 2006.

[57] J. D. Wines, R. Saitz, N. J. Horton, C. Lloyd-Travaglini, and J. H. Samet. Overdose after detoxification: A prospective study. *Drug and Alcohol Dependence*, 89:161–169, 2007.

Indices

Separate indices are provided for subject (concept or task) and for SAS commands. References to the HELP examples are denoted in *italics*.

Subject index

References to the HELP examples are denoted in *italics*.

SAS index

References to the HELP examples are denoted in *italics*.